Energy, Climate and the Environment

This is a critical time for nature-society relations as the global community faces the challenge of making food, energy, and transport systems more sustainable and resilient. Energy, climate and environmental systems are central to staying within planetary boundaries amid rapidly changing social, economic, political, and environmental conditions. This book series will focus on these topics, their interlocking impacts, interactions and uncertain outcomes. Contributions are encouraged from emerging and established scholars working within and/or across the energy, climate and environmental domains. The series provides an outlet for scholarship, debate, discussion, and dialogue between social sciences, humanities, physical and applied sciences. Its objective is to promote interdisciplinary and transdisciplinary research, whilst simultaneously fostering positive social change. The series provides an opportunity for scholars to connect state of the art science with the policymaking and practitioner community, accelerating the transition to a resilient future.

Nathanael Ojong

Solar Power Capitalism

How Green Energy Drains Bodies, Ecologies, and Futures

palgrave
macmillan

Nathanael Ojong
International Development Studies
York University
Toronto, ON, Canada

ISSN 2947-8561 ISSN 2947-857X (electronic)
Energy, Climate and the Environment
ISBN 978-3-032-13706-7 ISBN 978-3-032-13707-4 (eBook)
https://doi.org/10.1007/978-3-032-13707-4

Cover credit: Carolyn Franks

This Palgrave Macmillan imprint is published by the registered company Springer Nature Switzerland AG
The registered company address is: Gewerbestrasse 11, 6330 Cham, Switzerland

For Rita, Kezia, and Nathania

Preface

This book emerged from years of fieldwork across Ghana, Kenya, Tanzania, Malawi, and Cameroon, examining the contradictory geographies of solar power's promise. I began with a simple question: for whom is solar power a solution, under what conditions, and at what cost? The answer unfolded in the bauxite mines, the fenced lands of utility-scale solar plants, and the pay-as-you-go home systems in households. Each site revealed fragments of a larger circuitry linking extraction, finance, and everyday life.

As I followed these circuits, two concepts crystallized: insertion and depletion. Insertion captures the patterned incorporation of people, ecologies, and institutions into infrastructures of accumulation; depletion designates the bodily, ecological, and temporal exhaustion through which those infrastructures endure. Together they form a single circuit through which contemporary green capitalism reproduces itself. This conceptual pair grew from encounters with people in mining communities whose bodies bear the burdens of extracting green minerals, with women juggling solar repayments against food budgets, and with workers dismantling toxic solar waste.

The book's argument is unsettling: solar power capitalism thrives by converting hope into discipline and sustainability into exhaustion. It invites us to see solar energy not simply as a technical fix, but as a social relation shaped by inequality, gender, and finance. This book's guiding concern is with those whose everyday labor and endurance sustain the very infrastructures that promise their liberation. Their experiences reveal that genuine transformation requires confronting, not glossing over, the contradictions of the green transition.

Toronto, Canada — Nathanael Ojong

Acknowledgments Over the years that I have studied solar energy technologies in Africa, a journey culminating in the publication of this book, I have accumulated a number of debts to those in my social and professional networks. Without these networks of support, *Solar Power Capitalism* would have been less engaging. In the institutional sphere, the research was supported financially through first the Social Sciences and Humanities Research Council of Canada grant on 'The antinomies of renewable energy technologies: Debt, finance capitalism, and solar home systems' (#430-2021-00930). The anonymous reviewers appointed by Palgrave Macmillan were helpful in various ways. I gratefully acknowledge their helpful feedback but emphasize that this acknowledgment should not be read as their endorsement of the book.

Through their ideas, critiques, and questioning, several people urged me to think beyond the obvious, to probe more deeply and imagine more boldly what my research was revealing, including Deborah James, Elisabeth Prügl, Janet Roitman, Mohamed Sesay, and Philip Mader. Their feedback on the draft chapters contributed to improving the overall quality of this book. Of course, any mistakes that remain are mine alone. Some of the findings in this book were presented at the Royal Geographical Society (with Institute of British Geographers) Annual International Conference in London, England (2023); the European Association of Development Research and Training Institutes (EADI) General Conference in Lisbon, Portugal (2023); and the Energy and Society Network 6th International Joint Conference in Trento, Italy (2023). I thank the participants in the sessions where these findings were discussed for their thoughtful comments, which helped me refine several of the ideas developed in this book. Eyram Agbe and Annelise Gill-Wiehl were astute and hardworking collaborators with whom I co-authored three papers related to the broader project on solar electrification in Africa. I benefited from their ideas and suggestions. I am deeply grateful to the research assistants and participants in Cameroon, Ghana, Kenya, Malawi, and Tanzania whose generosity in sharing information made this book possible.

I wish to express my sincere appreciation to Rachael Ballard, editor at Palgrave Macmillan, for her steady support and productive collaboration throughout this project. I also thank Santhosh Kirubanithi and Cecile Schuetze-Gaukel for their excellent editorial assistance, which greatly facilitated the book's timely completion.

Most importantly, Rita, Kezia, and Nathania brought joy, laughter, and a wonderful sense of companionship to our home. It is to them that I owe the greatest debt of gratitude for their understanding of the long hours I devoted to work, including during my sabbatical.

Competing Interests The author has no competing interests to declare that are relevant to the content of this manuscript.

Contents

1 Engaging Solar Power Capitalism 1
1.1 Introduction 1
1.2 The Circuit of Insertion and Depletion 2
1.3 Methods and Cases 9
1.4 Structure of the Book 13
1.5 Concluding Remarks 15
References 15

2 Reworking Colonial Frontiers 23
2.1 Introduction 23
2.2 Colonial Grids 24
2.3 Reworking Empire 34
2.4 Concluding Remarks 40
References 40

3 Engineering Solar Energy Markets 45
3.1 Introduction 45
3.2 Redirecting Surplus Capital 47
3.3 Engineering Market Design 50

3.3.1 Period One (1993–2006): Pilot Projects 59
3.3.2 Period Two (2007–2016): Standards and Market Infrastructure 60
3.3.3 Period Three (2012 Onward): De-risking Private Capital 62
3.4 Statecraft in Dependence 73
3.5 Solar Power Boom and the Reproduction of Structural Dependence 79
3.6 Solar Power and the Reproduction of the Resource-Extraction-Export Model 87
3.6.1 A Market for Solar Power, but Not a Leading Manufacturer of Solar Goods 89
3.7 Concluding Remarks 93
References 94

4 Solar Finance and Everyday Extraction 107
4.1 Introduction 107
4.2 Intermediary Conduits 109
4.2.1 PAYG Platforms: Insertion Through Digital Leverage, Depletion Through Automated Darkness 109
4.2.2 Local SHS Suppliers and Installers: Insertion Through Sales, Depletion Through Scarce Incomes 120
4.2.3 Indigenous Financial Institutions: Insertion Through Collective Collateral, Depletion of Reciprocity 125
4.2.4 Peer Lenders: Insertion into Relational Debt, Depletion of Dignity and Trust 129
4.3 Circuits of Extraction: The Globalization of Insertion and Depletion 134
4.4 Concluding Remarks 138
References 138

5 Toxic Landscapes 145
5.1 Introduction 145
5.2 Ghana's Green Energy Triad: Bauxite Mine, Solar Plant, Scrapyard 146
5.2.1 Awaso Bauxite Mine 147
5.2.2 Kaleo-Lawra Utility-Scale Solar Plant 148
5.2.3 Agbogbloshie E-Waste Scrapyard 148
5.3 Embodied Violence 149
5.4 Toxic Afterlives: Insertion into E-Waste, Depletion Through Toxicity 157
5.4.1 Insertion into Hazardous Recycling, Depletion as Disposable Life 158
5.4.2 More-than-Human Bodies: Ecological Insertion and Depletion 161
5.5 Concluding Remarks 168
References 169

6 Gendered Infrastructures 173
6.1 Introduction 173
6.2 Governing Energy Through Gender 175
6.3 Enclosure, Time Taxes, and Patriarchal Spatial Orders 180
6.4 Toxic Work 185
6.5 Concluding Remarks 196
References 196

7 Circuits of Exhaustion 203
7.1 Confronting Contradictions 203
7.2 Reworking the Circuit 207
7.3 The Contradiction of Ambivalent Gains 210
7.4 From African Frontiers to Global Capitalism 211
7.5 Toward Alternative Futures 214
References 216

Index 221

About the Author

Dr. Nathanael Ojong is an associate professor in the Department of Social Science at York University in Canada. His research can be divided into three main streams. The first examines the effects of low-carbon electricity generation technologies (e.g., solar plants and solar home systems) in sub-Saharan Africa. The second explores how people make sense of credit and debt, as well as how they engage in borrowing and lending practices. His third research stream investigates the economic and sociocultural processes and conditions that shape entrepreneurship in the global South.

His work has been published in Energy Research & Social Science, Environmental Research Letters, Journal of Rural Studies, Social Science & Medicine, Journal of Business Research, and the International Journal of Entrepreneurial Behavior & Research. He is the author of *The Everyday Life of the Poor in Cameroon: The Role of Social Networks in Meeting Needs*, and the editor of *Off-Grid Solar Electrification in Africa*. He serves as an associate editor for *Sustainable Development* and as an associate editor for *Business Strategy and the Environment*.

Abbreviations

AfDB	African Development Bank
ALUCAM	Aluminum du Cameroun (Cameroonian Aluminum Company)
ARCO	Atlantic Richfield Oil Company
ASCA	Accumulating Savings and Credit Association
ASE	Ampyr Solar Europe
CFP	Compagnie Française des Pétroles
CSR	Corporate Social Responsibility
DC	Direct Current
DFID	Department for International Development
DFIs	Development Finance Institutions
DPGE	Development Plan General Electricity
DRC	Democratic Republic of the Congo
DRE	Decentralized Renewable Energy
EAQIP	Energy Access and Quality Improvement Project
EIA	Environmental Impact Assessment
EIB	European Investment Bank
ENELCAM	Énergie Électrique du Cameroun
EPC	Engineering–Procurement–Construction
ESMAP	Energy Sector Management Assistance Program

FiT	Feed-in Tariff
FOCAC	Forum on China–Africa Cooperation
FPUP	Federal Photovoltaics Utilization Program
GDP	Gross Domestic Product
GEDAP	Ghana Energy Development and Access Project
GEF	Global Environment Facility
GHS	Ghanaian Cedi
GOGLA	Global Off-Grid Lighting Association
IBRD	International Bank for Reconstruction and Development
IDA	International Development Association
IEA	International Energy Agency
IFC	International Finance Corporation
IMF	International Monetary Fund
IPO	Initial Public Offering
IRENA	International Renewable Energy Agency
KES	Kenyan Shilling
KWh	Kilowatt-hour
M2M	Machine-to-Machine
MIGA	Multilateral Investment Guarantee Agency
MoFA	Ministry of Food and Agriculture
MVA	Manufacturing Value Added
MWK	Malawian Kwacha
NECPA	National Energy Conservation Policy Act
NGO	Non-Governmental Organization
NGPA	Natural Gas Policy Act of 1978
NYSE	New York Stock Exchange
OECD	Organisation for Economic Co-operation and Development
OGS	Off-Grid Solar
OPEC	Organization of the Petroleum Exporting Countries
PAYG	Pay-As-You-Go
PPP	Public–Private Partnership
PURPA	Public Utilities Regulatory Policy Act
PV	Photovoltaic
RBF	Results-Based Financing
RoSCA	Rotating Savings and Credit Association
SDGs	Sustainable Development Goals
SEforALL	Sustainable Energy for All
SENELEC	Senegalese Electricity Distribution Company
SHS	Solar Home System

SIM	Subscriber Identity Module
TANESCO	Tanzania Electric Supply Company
TZS	Tanzanian Shilling
UNCTAD	United Nations Conference on Trade and Development
UNDP	United Nations Development Programme
USAID	United States Agency for International Development
VALCO	Volta Aluminum Company Limited
XAF	Central African CFA Franc

List of Figures

Fig. 3.1	Approved World Bank funding for off-grid goods and services (US$millions) (Data sourced from ESMAP 2017–2024)	65
Fig. 3.2	A self-reinforcing loop among sales, investment, and policy reforms in Africa's solar energy access markets	67
Fig. 3.3	Commodity dependence by region, 2019–2021 (Data sourced from UNCTADStat database. Available at: https://unctadstat.unctad.org/datacentre/ (accessed January 9, 2025))	81
Fig. 3.4	African countries where one product group was at least 50% of merchandise exports in 2023 (Data sourced from UNCTADStat database. Available at: https://unctadstat.unctad.org/datacentre/ (accessed January 9, 2025))	81
Fig. 3.5	Manufactured goods as share of merchandise exports, 2023 (Data sourced from UNCTADStat database. Available at: https://unctadstat.unctad.org/datacentre/ (accessed January 9, 2025))	83

Fig. 3.6 Manufacturing value added per capita at constant (2015) US$in Africa (Data sourced from UNIDO's Competitive Industrial Performance Index 2019–2021. Available at: https://stat.unido.org/analytical-tools/cip?country=124 (accessed January 9, 2025)) 85
Fig. 3.7 Export structure of Zambia and the Democratic Republic of Congo, 2023 (Data sourced from UNCTADStat database. Available at: https://unctadstat.unctad.org/datacentre/ (accessed January 9, 2025)) 90
Fig. 4.1 Solar panel on a roof, Kenya (Author's photo taken during fieldwork) 117
Fig. 4.2 Macro–micro circuits of financial accumulation 135
Fig. 5.1 Agbogbloshie e-waste scrapyard (Author's photo taken during fieldwork) 149
Fig. 5.2 Solar panel discarded at Agbogbloshie (Author's photo taken during fieldwork) 150
Fig. 5.3 Bauxite truck on unpaved Awaso road (Author's photo taken during fieldwork) 152
Fig. 5.4 Fenced Kaleo-Lawra solar plant (Author's photo taken during fieldwork) 157
Fig. 5.5 Worker dismantling a broken TV in Agbogbloshie (Author's photo taken during fieldwork) 159
Fig. 5.6 Polluted water body near Agbogbloshie (Author's photo taken during fieldwork) 166
Fig. 5.7 Tires arranged for e-waste burning in Agbogbloshie (Author's photo taken during fieldwork) 167

List of Tables

Table 1.1	Key indicators for the five country cases	12
Table 2.1	Dynamics of colonial electrification in five focal countries	27
Table 2.2	Core costs of producing a ton of aluminum in Ghana in 1960s	31
Table 3.1	OGS investment in the global South (excluding results-based financing)	61
Table 3.2	Escorting instruments and country examples	67
Table 3.3	Regional averages of manufacturing value added per capita at constant (2015) US$	84
Table 3.4	Leading importers of photovoltaic products in Africa	92
Table 7.1	Sites, mechanisms, and burdens in solar power capitalism	205

1

Engaging Solar Power Capitalism

1.1 Introduction

At dawn in Awaso, Ghana, trucks heavy with bauxite make their way down degraded roads, transporting ore that will be refined into the aluminum frames of solar panels. Thousands of miles away, in rural Kenya, a household's pay-as-you-go (PAYG) solar home system (SHS) is remotely shut off after missed mobile-money payments, leaving a family in sudden darkness.[1] These are not disconnected moments. They are linked fragments of the same circuitry: minerals extracted under conditions of dispossession, and households disciplined through financialized solar energy provision.

Viewed through what I later theorize as insertion and depletion, these scenes are not isolated instances but windows into the machinery of

[1] A PAYG SHS consists of one or more solar panels, a battery, charge controller, inverter (depending on system design), and bulbs (Sovacool & Griffiths, 2020), and uses a Global System for Mobile (GSM)/Machine-to-Machine (M2M) SIM and a software platform integrating mobile money to remotely monitor the system (Sanyal, 2017). Mobile money refers to a digital financial service that allows individuals to store, send, and receive money using their mobile phones—without needing access to a traditional bank account (Jack & Suri, 2011; Mas & Morawczynski, 2009; Maurer, 2012).

N. Ojong, *Solar Power Capitalism*, Energy, Climate and the Environment,
https://doi.org/10.1007/978-3-032-13707-4_1

accumulation that connects extraction, finance, and everyday life. I illustrate how such scenes of extraction and disconnection make visible the architecture of *solar power capitalism.*

I take solar power as an entry point, not the end point. The same dynamics that connect bauxite miners to global supply chains and bind Kenyan households to repayment schedules reappear in wind energy projects in India (Singh, 2025), in small hydropower projects in Vietnam (Dao, 2025), and forest-based carbon-offset initiatives in Latin America (Blanton et al., 2024; Cubas-Baez et al., 2025). The frontier of green capitalism is not confined to one technology or one region; it is an architecture of accumulation that displaces risk downstream and lifts value upstream across multiple energy forms. I use Africa's solar energy frontier as its starting point because it makes these dynamics unusually visible.[2] The continent's role in mineral extraction, off-grid solar electrification, and e-waste disposal provides a vantage for revealing how insertion and depletion organize both local livelihoods and global flows of green capital.[3]

The concepts developed here, insertion and depletion, travel well beyond Africa and well beyond solar power. They offer analytic tools for making sense of how contemporary green energy capitalism stabilizes itself through the incorporation of vulnerable actors and the exhaustion of the resources on which they rely.

1.2 The Circuit of Insertion and Depletion

Seen through the material landscapes of solar plants, SHSs, and solar waste, I argue that this advance of solar power across Africa is neither accidental nor the inevitable outcome of technological progress. It is

[2] For related arguments about how property and institutional architectures reproduce inequality in African resource economies, see Obeng-Odoom (2020).

[3] Off-grid solar electrification refers to the provision of electricity through standalone solar systems that operate independently of the central electricity grid, and includes solar photovoltaic goods such as solar lamps, lanterns, and mobile chargers—often in the capacity range of 1 to 5W—and solar home lighting systems ranging up to 100W (Few et al., 2019).

historically embedded, geographically selective, and politically orchestrated. This configuration reflects what Marx ([1867] 1990) and, later, Smith (2008) framed as uneven development—a 'seesaw' relocation of investment from stagnating, low-profit regions to fresh profit frontiers, leaving capital concentrated at one pole and labor surplus at the other.

Harvey (1982) sharpens this insight by showing how capital's ability to build up, and tear down, entire built environments, along with their social relations, yields differential returns to investors with varying time horizons. In moments of over-accumulation, surplus value is not only shifted spatially; it is increasingly absorbed by monetizing everyday life itself (Harvey, 1982; Langley, 2008). From this lens, Africa becomes doubly enticing: its vast terrain accommodates utility-scale solar hardware, while its loosely regulated financial architecture offers a laboratory for asset-backed lending and mobile-money intermediation (see Baker, 2023; Stock et al., 2023). The continent thus serves simultaneously as a frontier of solar power promise and a spatial fix for contradictions of capitalism generated elsewhere. What distinguishes Africa's solar power frontier, however, is how these dynamics assume a financialized form, made visible through insertion and depletion at the household and community level.

These processes, while financialized, are never solely economic. They are lived through bodies, households, and landscapes that absorb the pressures of accumulation.

I define *insertion* as the patterned incorporation of people, ecologies, and institutions into infrastructures of accumulation. It operates across colonial, financial, digital, and gendered registers, converting everyday labor, land, data, and care into stabilizers of global green value. Insertion is not reducible inclusion but a disciplinary process that makes life and environment investable.[4] *Depletion* denotes the bodily, ecological, and temporal exhaustion through which those infrastructures are sustained (see also Lingham & Johnston, 2025; Rai et al., 2014). It is not an aftermath of insertion but its counterpart, the slow wearing

[4] This book therefore joins a wider conversation on what Kaushal and Predmore (2025) term the 'women's inclusion agenda,' a transnational policy frame that celebrates women as prudent borrowers and market stabilizers even as it deepens financialization across very different geographies.

down that secures accumulation, turning the degradation of health, time, and ecosystems into the hidden subsidy of solar power capitalism. Both insertion and depletion form a single circuit through which contemporary green energy capitalism reproduces itself. These concepts form the analytic core of this book.

The analytical precision of these concepts becomes clearer when contrasted with existing frameworks. While I build on conceptual frames such as energy justice (Jenkins et al., 2021; Lacey-Barnacle et al., 2020; Sovacool & Dworkin, 2014), financialization (Bracking, 2020; Krippner, 2005; Langley, 2020; Mader et al., 2020), or uneven development (Omstedt & Ebner, 2024; Ortiz, 2025; Smith, 2008), I argue that they often miss how accumulation stabilizes itself through the simultaneous insertion of vulnerable actors and the depletion of their time, labor, and ecologies.[5] Insertion and depletion therefore sharpen our grasp of the contradictory dynamics of green energy, showing not only who is excluded but also how systemic inclusion itself becomes a mechanism of exhaustion. These conceptual lenses sharpen the picture, but the thread binding them together is insertion and depletion: incorporation into financialized circuits, and the exhaustion that renders those circuits investable.

Insertion and depletion move beyond regional specificity. While developed here through the African solar energy frontier, their utility is not confined to solar power. They travel across other spheres of green capitalism—lithium extractive activities in Latin America (Jerez et al., 2021; Kingsbury, 2023; Voskoboynik & Andreucci, 2022) and green extractivism in the United States (Dunlap et al., 2025) to wind park development in southern Mexico (Hesketh, 2022)—where risk is displaced along the margins and value accumulates at the apex of global circuits.[6]

[5] Energy justice focuses on the nexus between energy with respect to generation and delivery on the one hand, and justice on the other hand (Jenkins et al., 2016). It encompasses critical analyses of where (in)justice occurs within energy systems, including how justice might be attained (Jenkins et al., 2016; McCauley et al., 2013; Sovacool et al., 2017). Obeng-Odoom (2025) likewise urges a shift from ecological-economic to stratification-economic analyses of environmental justice.

[6] The scramble for cobalt in the Democratic Republic of Congo has displaced thousands of households (Baker, 2019).

From this angle, Africa's solar energy frontier offers a privileged vantage point for theorizing wider logics of twenty-first-century accumulation.

The stakes of these abstractions materialize in everyday life. In the case of PAYG SHSs, capital's search for new outlets does not simply build infrastructure; it entangles households in rolling repayment schemes where light depends on mobile-money transfers and remote shut-off switches. What is often framed as technological inclusion (Carr-Wilson & Sandeep, 2018; Yadav et al., 2019) rests on the conversion of precarious livelihoods into collateralized cash flows. Households pay in installments until 'ownership' is declared, but missed payments trigger disconnections and deepen spirals of dependency. Women stretch food budgets, negotiate informal loans, and regulate household consumption to keep payments flowing, through everyday acts that stabilize investor returns thousands of miles away.

Colonial energy regimes concentrated electricity in settler enclaves and administrative hubs, excluding rural populations and indigenous systems of energy provisioning (Hård, 2023; Marwah, 2017; Parsons, 2019; Worth, 1998). Today's solar frontier does not overturn that legacy; it reworks it. Policy packages backed by the World Bank and International Finance Corporation (IFC) portray Africa as an 'open frontier.' Profitability here hinges less on kilowatts generated than on securitized repayment streams. Household insertion into micro-debt regimes, and the associated depletion of marginalized people's labor and ecological resources, are the very conditions that render Africa's solar energy frontier investable.

To understand how this operates, we must follow how capital, policy, and technology fold everyday life into financial architectures.

Infrastructures that claim to democratize energy frequently entrench Africa's position as a site where surplus capital captures value while risk is shifted onto vulnerable populations. Territorial fixes (Harvey, 1982) enable financialization—'the increasing role of financial motives, markets and actors' in every infrastructural decision (Epstein, 2005: 3; see also Mawdsley, 2018). Scholars offer different emphases on this process. Krippner (2005) frames it as a pattern of accumulation where profits accrue primarily through financial channels rather than through trade or production. Bracking (2020) highlights its distinctiveness from commodification: a resource is commodified when extracted and sold, but it becomes financialized when it underpins an interest-bearing asset such as a carbon bond or biodiversity offset. Langley (2020) extends this to everyday life, showing how households are governed through financial logics of speculation, indebtedness, and risk.[7]

Financialization in turn disciplines space to behave like an asset. Subsequent chapters illustrate this spiral: from colonial exclusions (Chapter 2) through multilateral market-making (Chapter 3) to household repayment ecologies (Chapter 4), showing how insertion and depletion underpin solar power capitalism at every scale.

What these financial architectures obscure are the material and bodily forms of harm they reproduce. Spatial-financial fixes never arrive on neutral ground. They settle onto environments already scarred by resource booms, colonial grid exclusions, and gendered regimes of labor. Here, the concepts of structural and slow violence help us grasp how harm is embedded in infrastructures and dispersed over time.[8] Structural violence denotes harm that does not stem from an identifiable agent but is instead embedded within societal structures and manifests through systemic inequalities in power, resources, and life chances (Farmer, 2004; Galtung, 1969) while slow violence relates to attrition that is delayed, dispersed, and often illegible until bodies or ecologies reach a breaking point (Nixon, 2011).[9] What distinguishes the solar energy frontier is how these forms of violence are reproduced through two circuits: upstream extractive frontiers that supply minerals,

---

[7] For an overview of definitional debates, see Mader et al. (2020) 'Financialization: An introduction,' which synthesizes the field's macro, meso, and micro approaches.

[8] There is a link between structural and slow violence (Dao, 2025; Davies, 2022; Pain & Cahill, 2022), with the later framed as 'out of sight' and providing 'broader, more complex descriptive categories of violence enacted over time' (Nixon, 2011: 2–11). Slow violence should not be understood solely in terms of temporality or the protracted unfolding of harm; rather, it must also be situated within the structural asymmetries that enable such harm to quietly persist and deepen over time (Davies, 2022). The concept draws attention to the question of 'who gets to decide the course of human–environment relationships,' thereby foregrounding issues of power and agency (O'Lear, 2018: 94).

[9] Galtung argued that violence 'is present when human beings are being influenced so that their actual somatic and mental realizations are below their potential' (Galtung, 1969: 168), adding that 'structural violence is silent… [and] may be seen as about as natural as the air around us' (Galtung, 1969: 173).

and downstream repayment regimes that govern PAYG SHS households through automated shut-offs. Chapter 5 develops this argument by showing how extraction and enclosure in Ghana, together with the toxic afterlives of solar goods in Accra, exemplify the sacrifice zones of solar power capitalism.[10]

Insertion is the nexus that links financial architectures to land politics and everyday reproduction. But it does not operate in a vacuum; it is mediated through patriarchy, which determines whose time, credit, and mobility become elastic in the service of accumulation.

By patriarchy I refer a system of social structures and practices that institutionalizes men's dominance over women across family, community, workplace, and state (Walby, 1989, 2013). This shifts the focus from attitude to institutional control over women's labor, bodies, and mobility. Importantly, patriarchy does not appear in identical form everywhere. It takes shape through class, religion, and custom (Amone et al., 2021; Bawa, 2012; Dogo, 2014; Sanya, 2013). Hunnicutt (2009) extends this line of thinking with the notion of 'varieties of patriarchy,' emphasizing that it is ideology and structure at once, operating across macro, meso, and micro levels. Recent studies highlight how these dynamics are embodied in norms of masculinity and everyday practice (Bah & Barasa, 2023; Bradshaw et al., 2017).

Within these structures of power, women carve out spaces of agency, often through subtle negotiation rather than open defiance. Within such structures, women exercise agency, but they do so within sets of concrete constraints that create trade-offs (Gatwiri & Mumbi, 2016; Kandiyoti 1988). Kandiyoti's (1988) account of bargaining clarifies why women often pursue strategies that protect long-term security even when those strategies reaffirm elements of male authority. Patriarchal bargain captures the micro-politics through which women extract room for maneuver without confronting the entire edifice head-on (Kandiyoti, 1988; Kawarazuka et al., 2019). For instance, Kawarazuka et al. (2019) reveal how women in coastal Kenya mobilize contradictions within gendered arrangements of food provisioning to reinterpret roles and

[10] For a detailed discussion of the concept of sacrifice zones, see Juskus (2023).

incrementally shift responsibilities, that is, to bargain with patriarchy rather than directly overturn it.

One of the most understated yet potent strategies in this repertoire is silence. Silence can be wielded as a form of bargaining. It is not simply a sign of subordination. It can be a strategic stance that avoids escalation, protects dignity, and manages risk in settings where certain moves invite reprisal (Gatwiri & Mumbi, 2016).[11] In other words, silence functions as a situated practice shaped by particular forms of oppression and by the social, cultural, economic, political, and physical contexts in which women act (Kolawole, 1997). As a mechanism of bargaining with patriarchy, women may strategically acquiesce to patriarchal norms and gendered roles that appear to disadvantage them (Kandiyoti, 2005). Over time, however, these strategies do not simply manage risk; they exact a cumulative toll on women's bodies, time, and emotional reserves.

This juncture between survival and exhaustion is where the analysis of insertion and depletion converges most clearly. Such negotiations, however, are not without consequence. The very strategies that enable survival—strategic silence, calculated acquiescence, and everyday improvisation—also sustain the wider architecture of solar power capitalism. It is precisely here that depletion comes into view: the accumulated costs of insertion under patriarchal governance.

The conceptual payoff is twofold. First, centering patriarchal governance guards against romantic accounts of empowerment that conflate women's management of financial precarity with emancipation. Patriarchy is an institutional apparatus that organizes who decides, who pays, and whose labor underwrites stability.

A second implication follows directly from this structural insight. Recognizing depletion as constitutive clarifies why the same women who value evening light also report longer walks, tighter budgets, and riskier work. Contradiction is the condition of insertion under financialized solar energy provision.

---

[11] Silence can be power (Ryan-Flood & Gill, 2013). 'Silence can also be a means of resistance [...] It can also be a tool of transformation' (Stone, 2002: 19). Other feminists offer a different perspective on ways of addressing oppression. For instance, Thiam (1986) calls on African women to resist oppression through the act of voicing it.

1.3 Methods and Cases

The arguments advanced so far need grounding in lived realities. Insertion and depletion are dynamics that unfold across specific sites. For this reason, I turn to five country cases (Ghana, Kenya, Tanzania, Malawi, and Cameroon) that reveal the full circuitry of solar power capitalism.

The choice of cases reflects both regional diversity and conceptual purpose. Ghana exemplifies both upstream extraction (bauxite mining) and downstream waste (Agbogbloshie), while also hosting utility-scale solar plants that reproduce land enclosures. Kenya and Tanzania represent the most mature PAYG markets, where households are disciplined through mobile-money repayment regimes. Malawi and Cameroon highlight fragile incomes and the proliferation of informal solar energy markets. Collectively, these sites capture the range of mechanisms through which solar power capitalism is assembled: from extraction and finance to household debt and toxic afterlives of solar goods.

The analysis developed in this book draws on 349 semi-structured interviews with a wide range of participants, including users of SHSs, repairers of solar devices, e-waste workers, and residents of communities affected by solar plants and bauxite mining. Forty-four interviews were conducted in Malawi, 113 in Tanzania, 90 in Ghana, 57 in Cameroon, and 45 in Kenya. The research project was approved by the Human Participants Review Committee of York University (#2021-391/# e2022-344 /# e2023-376). Participants gave their consent to the publication of the results of this study. Fieldwork was complemented by textual analysis of corporate documents and reports from the World Bank and IFC on off-grid electrification and green finance. Rather than treating these materials as isolated sources, I read them relationally—as situated inscriptions within the circuitry of insertion and depletion that constitutes solar power capitalism. The result is an ethnographically-informed political economy that grounds theoretical claims in lived encounters and material landscapes.

The choice of cases also highlights geographical, economic, and institutional diversity. Cameroon lies in Central Africa, Tanzania and Kenya in East Africa, Ghana in West Africa, and Malawi in the south. Economically, Kenya and Cameroon are regional powerhouses, while Malawi ranks among the world's poorest nations, its economy heavily dependent on agriculture.[12] Varying levels of development shape both the approaches to solar electrification and the manifestations of solar power capitalism.

These contrasts reveal how global green capital interacts with local institutional contexts. The countries also embody distinct colonial energy legacies that continue to structure today's solar energy trajectories. Ghana, Kenya, and Malawi inherited British electrification schemes that privileged mines, settler townships, and export enclaves, bypassing rural areas. Tanzania's grid, shaped under German and later British rule, displayed the same urban bias, while Cameroon's dual French-British experience has left it intimately connected to the franc-zone monetary architecture, which centralizes part of its foreign reserves in the French treasury.[13] Despite these differences, all five countries have committed to expanding renewable energy, providing fertile ground for examining solar power capitalism in its multiple configurations.

Second, their solar electrification landscapes are heterogeneous. Kenya has a long history of off-grid solar electrification (Adwek et al., 2019; Jacobson, 2007), while Cameroon and Malawi remain in nascent stages. The PAYG model is pervasive in Kenya (Adwek et al., 2020; Carr-Wilson & Sandeep, 2018) and Tanzania (Ojong et al., 2025) but far less so in Malawi, Ghana, and Cameroon. In Malawi, people typically purchase SHSs upfront or one component at a time (Ojong et al., 2025; Samarakoon et al., 2021). Ghana's push toward utility-scale solar plants makes it particularly emblematic of the logics of dispossession embedded in green energy. These countries also differ in their regulatory environments and levels of international investment, illustrating the variegated governance of solar power capitalism.

Together, they capture the breadth of the solar value chain: upstream extraction (Ghana's bauxite mines and utility-scale plants); midstream consumer financialization (Kenya's and Tanzania's PAYG ecosystems of mobile money, indigenous financial institutions, and peer lending);

---

[12] https://www.worldbank.org/en/country/malawi/overview.

[13] For a detailed and critical discussion of the Franc zone and its political and economic implications, see 'The Franc Zone' (Barnay, 2025).

and downstream informal markets of counterfeit or component-based SHSs that typify Cameroon and Malawi, alongside disposal (Ghana's Agbogbloshie e-waste hub).[14]

Beyond structural diversity, gendered labor forms the connective tissue across these cases. The selection highlights how women's insertion sustains solar energy capitalism across institutional contexts. In Ghana's Kaleo-Lawra region, women walk farther for firewood as communal land is enclosed for the solar plant, absorbing the bodily toll that keeps investor returns appealing. In Kenya and Tanzania, the same women who championed SHS adoption must later orchestrate phased PAYG repayments, regulate household consumption, and negotiate informal loans, all while shouldering unpaid care work. Patriarchy here is an infrastructure of governance: it determines who decides, who pays, and whose labor underwrites solar power capitalism. These dynamics, illustrated through Chapters 4–6, reveal how gendered insertion and depletion stabilize green energy accumulation.

To ground these dynamics in concrete settings, I turn to country-specific cases, beginning with Malawi.

Malawi exemplifies how fragile incomes and informal markets sustain solar energy capitalism through volatility. Insertion manifests when households piece together SHSs component-by-component or buy second-hand units (Samarakoon et al., 2021), while depletion arises when women absorb the risks of informal loans amid harvest failures or medical crises. With one of the lowest grid connection rates globally (Table 1.1), families turn to solar energy not as infrastructure but as precarious survival, making Malawi vital for understanding how financial precarity underwrites off-grid solar expansion.[15]

Tanzania reveals how off-grid solar expansion depends on women's unpaid labor to sustain long repayment cycles. Households internalize insertion through the integration of daily survival into PAYG payment

[14] Eke et al. (2025) argue that financialization unevenly allocates risks and benefits between investors and local communities.

[15] Although matrilineal traditions persist in some regions in the country, patriarchy remains pervasive: women's land rights are typically mediated by male kin, and practices like wife inheritance curtail widows' autonomy (Tembo 2013; Tiessen 2008; Wilson & Kachipande 2020).

Table 1.1 Key indicators for the five country cases

Country	Population (million, 2024)	GDP (current US$ billion, 2024)	Electrification (%)	Off-grid solar product sales (July–Dec. 2023)
Malawi	21	11	15% overall (≈6% rural)	73,893 units
Tanzania	68	78	48% overall (≈28% rural)	137,000 units
Kenya	56	124	76% overall (≈68% rural)	1,181,746 units
Cameroon	29	51	72% overall (≈26% rural)	168,359 units
Ghana	34	82	89% overall (≈77% rural)	No national figure reported for H2 2023

Notes Data for national population, GDP, and electrification rates sourced from the World Bank's World Development Indicators database. Available at: https://data.worldbank.org/indicator (accessed September 21, 2025). Data for off-grid solar product sales sourced from GOGLA (2024)

schedules, while depletion surfaces in the stretching of women's time and credit as elastic resources. Patriarchal property norms further entrench these dynamics, concentrating land rights and decision-making in men's hands (see Buberwa, 2016; Pepinsky, 1992).

Kenya's mature PAYG ecosystem disciplines borrowers through mobile-money scoring, transforming electricity access into credit surveillance.[16] Women navigate this dense financial sphere by negotiating informal loans and regulating household consumption, exercising constrained agency within patriarchal architectures (see also Gatwiri & Mumbi, 2016; Kawarazuka et al., 2019). Here, insertion appears in the trade-offs that keep repayments flowing, while depletion is registered in the constant stretching of time, credit, and care to stabilize access.

In Cameroon, solar energy capitalism is anchored in credit intermediation. Access runs through microfinance and indigenous financial institutions that entangle households in repayment schedules, inserting

[16] Kenya remains Africa's largest market for off-grid solar kits, with 1.18 million units sold in the last half of 2023 alone (GOGLA, 2024).

women as borrowers while denying them equivalent rights over property or assets.[17] Depletion surfaces in the ways women's labor and credit stabilize household participation even as statutory equality remains weak and debt risks fall to those least protected.

Ghana reveals how utility-scale solar plants such as those at Kaleo and Lawra, reproduce colonial-era enclosures while deepening gendered burdens (see also Stock & Nyantakyi-Frimpong, 2025; Stock et al., 2023). Women absorb the bodily costs of land loss, walking farther for firewood and negotiating restricted access to commons as patriarchal governance narrows their bargaining power (see Chapter 6).[18] These scenes show how insertion and depletion anchor large-scale solar energy projects in everyday life.

A final clarification concerns the methodological stance: the case study approach was not chosen for comparative purposes. As noted earlier, the cases differ historically, economically, and politically, and these differences shape their solar energy landscapes in distinct ways. Yet what interests me are the parallels in how solar energy capitalism unfolds across them. Rather than comparing outcomes, I reveal processes in which transnational networks of capital and policy translate into localized circuits of insertion and depletion. Similarities therefore arise from shared entanglement in global green finance, while differences emerge through the mediating effects of geography, economy, and politics.

1.4 Structure of the Book

The chapters that follow build on these conceptual foundations by illustrating how insertion and depletion operate across Africa's solar energy frontier. Each chapter grounds the argument in specific sites, while also pointing outward to the wider architectures of green capitalism.

---

[17] In Cameroon, customary land laws and inheritance practices continue to limit women's ownership (Fonjong et al., 2012; Njieassam 2019).

[18] In Ghana, patriarchal kinship systems and cultural idioms continue to shape women's access to land and decision-making power (Ampofo & Atobrah, 2023; Dankwa 2018; Gyan et al., 2020).

Chapter 2, 'Reworking Colonial Frontiers,' examines the colonial legacies that structured energy access in Africa. In so doing, I show how electrification in settler enclaves and administrative towns produced enduring exclusions, and how contemporary solar energy projects reproduce these patterns. By situating solar power in a longer genealogy of colonial infrastructures, I highlight how the green energy transition inherits (and retools) earlier logics of dispossession evident across other contexts of energy transition.

Chapter 3, 'Engineering Solar Energy Markets,' turns to the role of multilateral lenders and development banks in developing Africa's solar energy markets. I demonstrate how concessional finance and policy blueprints positioned the continent as a laboratory for global capital. I argue that what is often framed as technical 'capacity-building' is, in fact, the creation of conditions under which private investors can securitize repayment streams. The chapter thus speaks not only to African solar power but to the global architecture of green finance, from carbon trading schemes to blended-finance models for wind and hydropower.

Chapter 4 moves into the household, showing how PAYG SHSs discipline everyday life through credit logics. Women, in particular, manage repayment schedules, stretch food budgets, and absorb reputational risks when light is cut off. What is framed as empowerment frequently entails insertion into fragile financial ecologies that stabilize accumulation elsewhere. While rooted in African case studies, the dynamics here resonate with how households in Latin America and Southeast Asia navigate the burdens of micro-debt tied to green energy technologies.

In Chapter 5, I examine solar power's extractive geographies. I follow the circuit from mineral frontiers in Ghana to the toxic afterlives of discarded solar devices in Accra. These sites illustrate how the promise of clean energy depends on sacrifice zones where bodies and ecologies absorb costs displaced from elsewhere. By situating solar energy alongside other extractive industries (for example, bauxite), I stress how depletion anchors green energy capitalism's supply chains globally.

Chapter 6, 'Gendered Infrastructures,' brings these dynamics back to questions of gendered agency. I illustrate how women negotiate patriarchal bargains and find ways to keep households afloat even as they are depleted by repayment regimes and hazardous work at e-waste sites. Instead of romanticizing this agency, I insist on its contradictory character: enabling survival while also securing capital flow.

The final chapter synthesizes the analysis into a single circuit of solar power capitalism. I situate Africa's solar power frontier within the wider global green energy economy, drawing comparative connections to India's solar plants and Latin America's lithium triangle. In doing so, I establish the analytic reach of insertion and depletion as tools for grasping how twenty-first-century capitalism organizes accumulation under the banner of sustainability. From mines to households to scrapyards, solar energy capitalism does not simply brighten lives; it drains bodies, ecologies, and futures.

1.5 Concluding Remarks

In this opening chapter I have laid the conceptual groundwork for understanding solar power capitalism through the twin lenses of insertion and depletion. I have shown that what appears as a discourse of technological progress is, in fact, an architecture that incorporates vulnerable actors into financialized circuits while exhausting their time, bodies, and ecologies. In demonstrating how Africa's solar energy frontier fuses colonial design logics with contemporary financialization, I have argued that insertion and depletion are constitutive mechanisms of green accumulation. With this conceptual scaffolding in place, I now turn to the historical blueprints that made such architectures possible, showing how colonial grids and concessional finance created the very infrastructures through which insertion and depletion would later operate.

References

Adwek, G., Boxiong, S., Arowo, M., Ndolo, P., & Shimmon, J. (2019). Review of solar energy development in Kenya: Opportunities and challenges. *Renewable Energy Focus, 29*, 123–140.

Adwek, G., Shen, B., Ndolo, P. O., Siagi, Z. O., Chebet, C., Kemunto, C. M., & Patrobers, S. (2020). The solar energy access in Kenya: A review focusing on Pay-As-You-Go solar home system. *Environment, Development and Sustainability, 22*(5), 3897–3938.

Amone, C., Bond, J., Chiweshe, M. K., Fusari, V., Malunga, B., Mandhu, F., & Siang'andu, E. T. (2021). *Patriarchy and gender in Africa*. Rowman & Littlefield.

Ampofo, A. A., & Atobrah, D. (2023). Women in Ghana. In *Oxford research encyclopedia of African history*. https://doi.org/10.1093/acrefore/9780190277734.013.539

Bah, A. B., & Barasa, M. N. (2023). Indigenous knowledge and the social construction of patriarchy: The case of the Bukusu of Kenya. *Critical Sociology, 49*(2), 217–232.

Baker, G. (2019, December 19). Thousands to lose homes after £75bn cobalt discovery in Democratic Republic of Congo. *The Times*. https://www.thetimes.co.uk/article/thousands-to-lose-homes-after-75bn-cobalt-discovery-in-democratic-republic-of-congo-0t58pn6h7. Accessed November 5, 2024.

Baker, L. (2023). New frontiers of electricity capital: Energy access in sub-Saharan Africa. *New Political Economy, 28*(2), 206–222.

Barnay, M. (2025, January 10). The Franc Zone. *New Left Review: Sidecar*.

Bawa, S. (2012). Women's rights and culture in Africa: A dialogue with global patriarchal traditions. *Canadian Journal of Development Studies/Revue canadienne d'études du développement, 33*(1), 90–105.

Blanton, A., Mohan, M., Galgamuwa, G. P., Watt, M. S., Montenegro, J. F., Mills, F., & Ewane, E. B. (2024). The status of forest carbon markets in Latin America. *Journal of Environmental Management, 352*, 119921.

Bracking, S. (2020). Financialization and the environmental frontier. In *The Routledge international handbook of financialization* (pp. 213–223). Routledge.

Bradshaw, S., Linneker, B., & Overton, L. (2017). Extractive industries as sites of supernormal profits and supernormal patriarchy? *Gender & Development, 25*(3), 439–454.

Buberwa, D. K. (2016). Patriarchy ideology and workload inequality within household. *Journal of Human-Social Science, 16*(1), 1–9.

Carr-Wilson, S., & Sandeep, P. (2018). Pay-as-you-go: How a business model is helping light millions of rural Kenyan homes with solar. *Case Studies in the Environment, 2*(1), 1–10.

Cubas-Baez, A., Sunderlin, W., Larson, A. M., & Frechette, A. (2025). From promises to practice: Persistent challenges in safeguarding local rights in forest carbon market initiatives. *CABI Reviews, 20*(1), 0025.

Dankwa, S. (2018). Culture of family ideals and perceived subjugating positions of women in patriarchy society: The way forward. *Culture* (Québec), *8*(24).

Dao, N. (2025). Small hydropower, slow violence, and gendered struggles in Northwest Vietnam. *Environment and Planning C: Politics and Space*. https://doi.org/10.1177/23996544251318589

Davies, T. (2022). Slow violence and toxic geographies: "Out of sight" to whom? *Environment and Planning c: Politics and Space, 40*(2), 409–427.

Dogo, S. A. (2014). The Nigerian patriarchy: When and how. *Cultural and Religious Studies, 2*(5), 263–275.

Dunlap, A. A., Novaković, B., & Sovacool, B. K. (2025). Mining as environmentalism: Green/grey extractivism and the production of extractive subjectivities around the Rio Tinto Kennecott mine in the United States. *Globalizations*, 1–36.

Eke, S., Grant, J. A., Mayanja, E., & Andrews, N. (2025). Transnational capital and the scramble for land and profit: Financialization, agrarian development, and resource conflict in Africa. *World Development, 194*, Article 107076.

Epstein, G. A. (Ed.). (2005). *Financialization and the world economy*. Edward Elgar.

Farmer, P. (2004). An anthropology of structural violence. *Current Anthropology, 45*(3), 305–325.

Few, S., Schmidt, O., & Gambhir, A. (2019). Energy access through electricity storage: Insights from technology providers and market enablers. *Energy for Sustainable Development, 48*, 1–10.

Fonjong, L., Sama-Lang, I. F., & Fombe, L. F. (2012). Implications of customary practices on gender discrimination in land ownership in Cameroon. *Ethics and Social Welfare, 6*(3), 260–274.

Galtung, J. (1969). Violence, peace, and peace research. *Journal of Peace Research, 6*(3), 167–191.

Gatwiri, G. J., & Mumbi, K. A. (2016). Silence as power: Women bargaining with patriarchy in Kenya. *Social Alternatives, 35*(1), 13–18.

GOGLA. (2024). *Global off-grid solar market report: Semi-annual sales & impact data (Jul–Dec 2023)*. https://www.gogla.org/wp-content/uploads/2024/05/GOGLA_Sales-and-Impact-ReportH2-2023_FINAL.pdf. Accessed May 7, 2024.

Gyan, C., Abbey, E., & Baffoe, M. (2020). Proverbs and patriarchy: Analysis of linguistic prejudice and representation of women in traditional Akan communities of Ghana. *Social Sciences, 9*(3), 22.

Hård, M. (2023). Accessing electricity in East Africa: Dar es Salaam dwellers pursue power. In *Microhistories of technology* (pp. 101–128). Springer.

Harvey, D. (1982). *The limits to capital*. University of Chicago Press.

Hesketh, C. (2022). Clean development or the development of dispossession? The political economy of wind parks in Southern Mexico. *Environment and Planning e: Nature and Space, 5*(2), 543–565.

Hunnicutt, G. (2009). Varieties of patriarchy and violence against women: Resurrecting "patriarchy" as a theoretical tool. *Violence against Women, 15*(5), 553–573.

Jack, W., & Suri, T. (2011). *Mobile money: The economics of M-PESA (No. w16721)*. National Bureau of Economic Research.

Jacobson, A. (2007). Connective power: Solar electrification and social change in Kenya. *World Development, 35*(1), 144–162.

Jenkins, K., McCauley, D., Heffron, R., Stephan, H., & Rehner, R. (2016). Energy justice: A conceptual review. *Energy Research & Social Science, 11*, 174–182.

Jenkins, K., Sovacool, B. K., Mouter, N., Hacking, N., Burns, M.-K., & McCauley, D. (2021). The methodologies, geographies, and technologies of energy justice: A systematic and comprehensive review. *Environmental Research Letters, 16*, Article 043009.

Jerez, B., Garcés, I., & Torres, R. (2021). Lithium extractivism and water injustices in the Salar de Atacama, Chile: The colonial shadow of green electromobility. *Political Geography, 87*, Article 102382.

Juskus, R. (2023). Sacrifice zones: A genealogy and analysis of an environmental justice concept. *Environmental Humanities, 15*(1), 3–24.

Kandiyoti, D. (1988). Bargaining with patriarchy. *Gender & Society, 2*(3), 274–290.

Kaushal, T., & Predmore, S. (2025). The women's inclusion agenda: Gender and everyday practices across registers of finance. *Finance and Society, 11*(1), 81–102.

Kawarazuka, N., Locke, C., & Seeley, J. (2019). Women bargaining with patriarchy in coastal Kenya: Contradictions, creative agency and food provisioning. *Gender, Place & Culture, 26*(3), 384–404.

Kingsbury, D. V. (2023). Lithium's buzz: Extractivism between booms in Bolivia, Argentina, and Chile. *Cultural Studies, 37*(4), 580–604.

Krippner, G. R. (2005). The financialization of the American economy. *Socio-Economic Review, 3*(2), 173–208.

Lacey-Barnacle, M., Robison, R., & Foulds, C. (2020). Energy justice in the developing world: A review of theoretical frameworks, key research themes and policy implications. *Energy for Sustainable Development, 55*, 122–138.

Langley, P. (2008). *The everyday life of global finance: Saving and borrowing in Anglo-America*. Oxford University Press.

Langley, P. (2020). The financialization of life. In *The Routledge international handbook of financialization* (pp. 68–78). Routledge.

Lingham, J. T., & Johnston, M. (2025). Running on empty: Depletion and social reproduction in Myanmar and Sri Lanka. *Antipode, 57*(2), 494–514.

Mader, P., Mertens, D., & Van Der Zwan, N. (2020). Financialization: An introduction. In *The Routledge international handbook of financialization* (pp. 1–16). Routledge.

Marwah, H. (2017). Electricity access inequality in sub-Saharan Africa, 1950–2000. *African Economic History, 45*(2), 113–144.

Marx, K. (1990). *Capital* (Vol. 1). Penguin Classics.

Mas, I., & Morawczynski, O. (2009). Designing mobile money services: Lessons from M-PESA. *Innovations, 4*(2), 77–91.

Maurer, B. (2012). Mobile money: Communication, consumption and change in the payments space. *Journal of Development Studies, 48*(5), 589–604.

Mawdsley, E. (2018). Development geography II: Financialization. *Progress in Human Geography, 42*(2), 264–274.

McCauley, D. A., Heffron, R. J., Stephan, H., & Jenkins, K. (2013). Advancing energy justice: The triumvirate of tenets. *International Energy Law Review, 32*(3), 107–110.

Nixon, R. (2011). *Slow violence and the environmentalism of the poor*. Harvard University Press.

Njieassam, E. E. (2019). Gender inequality and land rights: The situation of indigenous women in Cameroon. *Potchefstroom Electronic Law Journal/Potchefstroomse Elektroniese Regsblad, 22*(1).

Obeng-Odoom, F. (2020). *Property, institutions, and social stratification in Africa*. Cambridge University Press.

Obeng-Odoom, F. (2025). Ecology and power: From environmental and ecological economics to stratification economics. *Review of Evolutionary Political Economy*, 1–16.

Ojong, N., Gill-Wiehl, A., & Agbe, E. (2025). Cash, credit, or kin? financing pathways and the uneven sustainability of off-grid solar electrification in

Tanzania and Malawi. *Sustainable Development*. https://doi.org/10.1002/sd.70009

O'Lear, S. (2018). *Environmental geopolitics*. Rowman & Littlefield.

Omstedt, M., & Ebner, N. (2024). Introduction: Uneven development and social difference in capitalism. *Environment and Planning a: Economy and Space, 56*(4), 1298–1303.

Ortiz, R. J. (2025). Spatial competition and unevenness in global capitalism: Labor, nature, and the seesaw of capital. *Critical Sociology, 51*(2), 283–300.

Pain, R., & Cahill, C. (2022). Critical political geographies of slow violence and resistance. *Environment and Planning c: Politics and Space, 40*(2), 359–372.

Parsons, T. (2019). *Urban development and the East African railways*. Oxford University Press.

Pepinsky, H. E. (1992). Corruption, bribery and patriarchy in Tanzania. *Crime, Law and Social Change, 17*(1), 25–51.

Rai, S. M., Hoskyns, C., & Thomas, D. (2014). Depletion: The cost of social reproduction. *International Feminist Journal of Politics, 16*(1), 86–105.

Ryan-Flood, R., & Gill, R. (Eds.). (2013). *Secrecy and Silence in the Research Process: Feminist Reflections*. Routledge.

Samarakoon, S., Bartlett, A., & Munro, P. (2021). Somewhat original: Energy ethics in Malawi's off-grid solar market. *Environmental Sociology, 7*(3), 164–175.

Sanya, B. N. (2013). Disrupting patriarchy: An examination of the role of e-technologies in rural Kenya. *Feminist Africa, 18*, 12–24.

Sanyal, S. (2017). *"Pay-As-You-Go" solar could electrify rural Africa*. World Resources Institute. https://www.wri.org/insights/pay-you-go-solar-could-electrify-rural-africa. Accessed September 15, 2025.

Singh, D. (2025). A new regime of dispossession in neo-liberal India? wind energy, hindutva, and land politics in western Gujarat. *Journal of Contemporary Asia, 55*(3), 496–517.

Smith, N. (2008). *Uneven development: Nature, capital, and the production of space*. University of Georgia Press.

Sovacool, B. K., & Griffiths, S. (2020). The cultural barriers to a low-carbon future: A review of six mobility and energy transitions across 28 countries. *Renewable and Sustainable Energy Reviews, 119*, Article 109569.

Sovacool, B. K., Burke, M., Baker, L., Kotikalapudi, C., & Wlokas, H. (2017). New frontiers and conceptual framework for energy justice. *Energy Policy, 105*, 677–691.

Sovacool, B. K., & Dworkin, M. H. (2014). *Global energy justice*. Cambridge University Press.

Stock, R., & Nyantakyi-Frimpong, H. (2025). Laboring for light: Energy unfreedoms and freedom dreams of solar labor in Ghana. *The Journal of Peasant Studies, 52*(2), 235–257.

Stock, R., Nyantakyi-Frimpong, H., Antwi-Agyei, P., & Yeleliere, E. (2023). Volta photovoltaics: Ruptures in resource access as gendered injustices for solar energy in Ghana. *Energy Research & Social Science, 103*, Article 103222.

Stone, E. (2002). *Rising from deep places: Women's lives and the ecology of voice and silence*. Peter Lang Publishing Inc.

Tembo, M. J. (2013). *Gender based structu [r] al violence in relation to the traditional practice of wife inheritance: The case of Malawi: An empirical study of the violence experienced by widows involved in wife inheritance practice* (Master's thesis). Universitetet i Nordland.

Thiam, A. (1986). *Speak out, black sisters: Feminism and oppression in black Africa*. Pluto Press London.

Tiessen, R. (2008). Small victories but slow progress: An examination of women in politics in Malawi. *International Feminist Journal of Politics, 10*(2), 198–215.

Voskoboynik, D. M., & Andreucci, D. (2022). Greening extractivism: Environmental discourses and resource governance in the "Lithium Triangle." *Environment and Planning e: Nature and Space, 5*(2), 787–809.

Walby, S. (1989). Theorising patriarchy. *Sociology, 23*(2), 213–234.

Walby, S. (2013). *Patriarchy at work: Patriarchal and capitalist relations in employment, 1800–1984*. John Wiley & Sons.

Wilson, A., & Kachipande, S. (2020). Women in Malawi. In *Oxford research encyclopedia of African history*. https://doi.org/10.1093/acrefore/9780190277734.013.694

Worth, D. (1998, September 4–5). *Lighting the darkness: The gas supply industry in nineteenth century Cape Town, and its role in the formation of an industrial landscape*. Paper presented at the Association for Industrial Archaeology Annual Conference, North Abbot, Devon.

Yadav, P., Heynen, A. P., & Palit, D. (2019). Pay-As-You-Go financing: A model for viable and widespread deployment of solar home systems in rural India. *Energy for Sustainable Development, 48*, 139–153.

2

Reworking Colonial Frontiers

2.1 Introduction

Energy systems in Africa, I argue, have never been neutral; from their colonial inception they underpinned political power, economic agendas, and imperial domination. Drawing on Harvey's (1982) concept of the spatial fix and scholarship on infrastructural power (Pinzur, 2021; Roitman, 2025), I show that Africa's solar energy boom is less a rupture than a reprise of a century-long project to territorialize electricity around imperial and, later, transnational capital (see also McDonald, 2012). I identify three design logics that recur across the cases: enclave privilege, in which electricity consolidates power around settler and industrial nodes; externalization of financial risk, where the costs of infrastructure are shifted onto local populations or states; and concessional underwriting, in which loans and aid embed external managerial control. I trace these dynamics in three ways: colonial grids that privileged mines and export corridors; their postcolonial retooling through dams and smelters; and the persistence of concessional financing that foreclosed energy sovereignty. These logics are best understood as modalities of insertion and depletion.

N. Ojong, *Solar Power Capitalism*, Energy, Climate and the Environment, https://doi.org/10.1007/978-3-032-13707-4_2

In advancing this argument, the chapter also speaks to two broader literatures. Energy justice scholarship emphasizes distributional, procedural, and recognitional dimensions of access (Jenkins et al., 2016; Ojong, 2022; Sovacool & Dworkin, 2015), yet my analysis shows how these injustices are outcomes designed into infrastructure itself.[1] At the same time, infrastructure studies have highlighted the political agency of networks, grids, and hydro-complexes (Folkers, 2024; Pinzur, 2021; Roitman, 2025); here I extend that insight by demonstrating how Africa's electrification projects were structured to reproduce global hierarchies of capital.

By situating solar electrification within the historical continuum of colonial and postcolonial capitalism, I show that engineered scarcity and selective provision were deliberate design features of energy systems. These features endure through the twin dynamics of insertion and depletion: insertion compels African households into precarious energy bargains, while depletion describes the ecological, bodily, and temporal costs that stabilize accumulation. In short, I use these dynamics to reveal how colonial design logics persist in contemporary solar power capitalism.

2.2 Colonial Grids

The genesis of electrification in Africa reveals an enduring colonial legacy, where electrification policies primarily served imperialist economic and political agendas. Electricity was introduced to the African continent during the 1880s and 1890s via colonial administrators and private companies (Showers, 2011). Under colonial regimes, the deployment of energy systems was not driven by a commitment to holistic development

[1] 'Distributive justice' focuses on inequities in the distribution of benefits and harms across an energy system (Jenkins et al., 2016; Walker & Day, 2012) and assesses where 'questions about the desirability of technologies in principle become entangled with issues that relate to specific localities' (Owens & Driffill, 2008: 4414). 'Recognition Justice' considers people whose views are side-lined in an energy system and how they should be recognized (Jenkins et al., 2016). 'Procedural justice' centers on whether decision-making processes regarding energy systems are fair (Jenkins et al., 2016; McCauley, 2018; Yenneti & Day, 2015).

but was instead an extension of capitalist accumulation strategies. Electrification prioritized export-oriented industrial hubs, mining, and settler communities, leaving domestic and rural needs marginalized (Hård, 2023; Marwah, 2017; Parsons, 2019; Worth, 1998). As Showers (2011: 195) accurately frames it, 'in most of colonial Africa, electricity was not seen as important for African or non-urban settlers' domestic lives.'

Across colonies, the same logic of selective provision reappeared through concessionary arrangements that privileged urban centers and industrial enclaves. Several colonial governments adopted a laissez-faire policy, granting concessions to private enterprises to develop electricity systems.[2] Electrification was confined to urban centers and areas of economic significance to colonial administrations (Parsons, 2019; van der Straeten, 2021). For instance, electricity was channeled to gold mining activities in South Africa (Hausman et al., 2008). In British-controlled Kenya, electrification efforts were largely concentrated in Nairobi and Mombasa, bypassing rural regions where the majority of the population resided (van der Straeten, 2021). Similarly, in Tanganyika (now Tanzania), British policies systematically ignored rural energy needs while fostering industrial activity in urban centers (Mchome, 2022). Kenya's case exemplifies how electrification was channeled to settler towns, while Tanganyika shows the same logic at work in a rail-and-port economy. Both highlight how rural African communities were systematically excluded.

These exclusions reflected an ideology of efficiency that equated economic productivity with settler and industrial demand. That same rationale reappears today in the language of financial risk deployed by PAYG solar firms. Just as Tanganyika officials once justified urban electrification on the grounds of efficiency, some of today's PAYG solar

[2] Before the turn of the twentieth century, a private German company provided electric street lighting to Dar es Salaam, the administrative center of German East Africa (now Tanzania), using wood-fired steam turbines (Byatt, 1920). In 1902, the first electricity plant in the French Protectorate of Tunisia was constructed at La Goulette by a concessionaire who also supplied gas to Tunis (Showers, 2011). More companies later built power generation plants in Sousse (1905), Sfax (1907), Ferryville (1909), and Bizerte (1911) (Cecelski et al., 2005). In the British Gold Coast (now Ghana), electricity was initially supplied through diesel generators that were installed, owned, and operated by industrial facilities such as mines and factories, along with other businesses and municipalities (RCEER, 2005).

firms justify exclusion of certain segments of the population by citing repayment risk (see also Baker, 2023). The resonance between colonial records and contemporary narratives reveals how old rationales are repackaged under the language of modern finance. Enclave privilege and selective provision rampant in the colonial epoch re-emerges in today's off-grid solar frontier, where PAYG discipline access through weekly, biweekly, or monthly repayments and the threat of lockout. As I argue in Chapter 4, these devices recode colonial logics into financial architectures that render household liquidity, and especially women's time and credit, the guarantors of energy continuity.

Enclave privilege visible in the colonial era re-emerges in contemporary society as insertion: incorporation into selective grids and repayment regimes. Depletion is the afterlife, time lost, ecologies degraded, and reproductive labor stretched to sustain infrastructures never designed for them.

In a similar vein, the Edea hydroelectric dam, the largest built in French Africa during the 1950s, was constructed on the lower Sanaga River in Cameroon (Lasserre, 1958) to serve the long-term needs of colonial and postcolonial capitalists. The dam met the needs of French business enterprises, especially those located in Douala, the industrial hub (Table 2.1). These cases illustrate the first recurring design logic of solar power capitalism: enclave privilege, where electrification consolidates power around narrow industrial and settler enclaves rather than serving the wider population.

The grid's geography codified colonial profitability rather than public utility, highlighting how infrastructure was engineered to channel returns to imperial industry rather than expand access for African households. This logic reappears in Chapter 4, where most PAYG solar firms privilege solvent households and marginalize others. Enclave privilege, in other words, is a spatial expression of insertion: it incorporates certain households or firms while excluding others, leaving depletion to accumulate in the lives and ecologies left outside the circuit.

Beyond spatial exclusion, colonial electrification projects displaced risks onto ecologies and communities. Colonial hydroelectric dams also triggered far-reaching environmental externalities, for example, catchment-area deforestation, loss of farms, etc. as well as displacements

Table 2.1 Dynamics of colonial electrification in five focal countries

Country	Colonial power at time of first supply	Initial electrification	Principal beneficiaries (colonial era)	Dominant financing model
Ghana	Britain	Diesel generators serving Gold Coast mines and municipal factories	Gold and manganese mining enclaves; settler municipalities in Accra and Sekondi	Private industrial self-generation that later segued into state-planned hydro (Volta) backed by UK/IBRD loans
Kenya	Britain	Nairobi municipal supply (extended to Mombasa by 1913)	White settler townships, railways, colonial civil service	30-year concession to East African Power & Lighting
Tanzania (then German East Africa/Tanganyika)	Germany and later Britain	First electricity-generating plant in Dar es Salaam, established by the Berlin-based East African Railroad Company in 1905 (Hård, 2023)	Company's own needs, railways, and harbor (Hård, 2023)	Private capital financing
Cameroon	Germany (pre-1916) France	Edéa hydroelectric complex	Aluminum plant smelter; industrial corridor	French colonial state loans (Caisse centrale) & ENELCAM/ALUCAM joint venture
Malawi (then Nyasaland)	Britain	Diesel generators	Colonial administration and white industrialists/landowners of tea and tobacco plantations	Private and commercial investment, municipal initiatives

Sources Hård (2023), Laparra (2002), Lasserre (1958), Mchome (2022), Showers (2011) and van der Straeten (2021)

(Baxter, 1977; Obeng, 1977; Showers, 2011). These forced displacements and ecological burdens exemplify the second design logic: the externalization of financial and social risk. Costs of construction and operation were shifted onto local ecologies and communities, while the benefits of electrification accrued to colonial firms. This logic of exporting risk away from investors onto African households recurs in later hydroelectric and solar projects. In Chapter 5, I illustrate how this logic persists in the toxic afterlives of solar waste. Here, externalization of risk reveals how depletion stabilizes accumulation: ecological and bodily exhaustion are made invisible so that infrastructure appears functional. It is this enduring logic of depletion that Chapter 5 makes explicit.

Malawi's reliance on diesel generation for plantation economies and Tanzania's donor-driven hydroelectric expansion both show how even modest or externally—financed systems entrenched enclave logics. These cases make clear that dependency and exclusion were continent-wide, not limited to larger colonial economies like Ghana or Kenya.

These spatial and ecological asymmetries were reinforced by symbolic displays of technological power. Early electrification projects, such as the installation of electric streetlights in settler cities like Kimberley, South Africa, in 1882, were symbolic gestures of technological superiority (Showers, 2011). Similar dynamics unfolded in West Africa: in Lagos Colony, the indigenous populations themselves lobbied for and ultimately secured electric street-lighting in 1898, demonstrating that in some contexts, local actors were anything but passive recipients of colonial infrastructure (Adebayo, 2024).

In Dar es Salaam, debates over installing electric lampposts highlighted the uneven prioritization of colonial urban spaces over indigenous needs (Hard, 2023). These projects prioritized settler communities and administrative centers, leaving indigenous populations without access to energy services.

But colonial electrification's exclusions were not only spatial or economic; they were also profoundly gendered. A striking omission in most colonial energy histories is how electrification reinscribed gendered divisions of labor. In settler towns, reliable lighting increased male industrial shifts, while the absence of power in rural areas consigned women to fuelwood collection and domestic work. Hydropower schemes such as

Cameroon's Edéa hydroelectric dam displaced farming communities in which women were primary food producers, compounding their unpaid reproductive load (Zébazé Togouet et al., 2023). By foregrounding these silent burdens, I lay the groundwork for Chapter 6's argument that solar energy reorganizes women's insertion under patriarchal governance even as it celebrates empowerment.

These burdens resonate with the domestic management dilemmas I later reveal in PAYG SHS households, clarifying the continuity between colonial energy exclusions and women's insertion into solar power capitalism (see Chapter 6). From this perspective, the displacement of women's labor into unpaid reproductive work foreshadows what I later theorize as insertion: the conversion of gendered responsibilities into hidden infrastructure for capital accumulation. Chapter 6 returns to this continuity explicitly.

Furthermore, colonial electrification infrastructures also reflected broader imperialist ambitions. This was particularly the case with hydroelectric projects. Electricity generation served European economic interests, particularly mining and agriculture, which were pivotal to the colonial economies. For example, hydropower development in Southern Rhodesia (now Zimbabwe) and Northern Rhodesia (now Zambia) was directly tied to the extraction of copper and other minerals, revealing the colonial prioritization of resource exploitation over local development (Marwah, 2017; Showers, 2011). In Nigeria, most of the electricity produced by the Nigeria Electricity Supply Company (NESCO), which acquired the existing Kwa station and established a hydroelectric system at Kurra Falls in 1929, was primarily utilized by the tin mines and their concentrating plant (Simpson, 1969). Although the Akosombo dam in Ghana was constructed after the country gained independence, initial studies were carried out when the country was still under British rule. The primary goal of the British was to use power generated from the dam for the processing of aluminum (Adu-Gyamfi et al., 2020). Colonial Britian had argued that supplying power from the Volta to smaller towns was inefficient and uneconomical due to their limited industrial and economic demand (DPGE, 1954).

The economic rationale behind these infrastructures was equally imperial. The design of the Edéa hydroelectric in French Cameroon included an industrial center for aluminum, prefiguring what the major facilities planned will be (Lasserre, 1958). In 1954 was founded the Cameroonian Aluminum Company (ALUCAM), a subsidiary of two major French producers, Péchiney and Ugine (Lasserre, 1958). The construction of the aluminum plant, about 600 meters from the Edea power plant, started in January 1955 and was completed in 1958.[3] For its part, *Energie électrique du Cameroun* (ENELCAM), established in 1948, invested XAF 3.5 billion [US$6,216,633] for the extension of the hydroelectric plant (Edea II) to meet an increase in demand for electricity (Laparra, 2002), especially for colonial economic needs. ALUCAM's financing was ensured by capital increases to XAF10 billion [US$17,764,690] and a loan of XAF6 billion [US$10,660,515] from the Caisse centrale over 20 years.[4] On its part, ENELCAM financing of the extension (Edéa II) included a loan of XAF2.3 billion [US$4,086,653] from the Caisse centrale to be repaid over forty years (Laparra, 2002). At the beginning, the alumina was imported from France while awaiting the commissioning of the Guinean Fria factory, which went operational in 1960 (Laparra, 2002).[5] The channeling of hydroelectric capacity into ALUCAM and allied facilities came at the expense of electrifying Cameroonian towns and agrarian economies, locking development into an enclave model and stunting domestic diversification.

It is worth noting that just like the construction of the Edea hydroelectric dam was not done because it was beneficial to the colony, Cameroon, the decision by French political and economic elites to construct the aluminum processing plant in the country can be understood through a combination of economic, geopolitical, and structural factors rooted in the dynamics of colonialism. First, the proximity of the plants to raw

---

3 The project estimate, drawn up at the end of 1954, included the entire factory (electric line and substation, electrolysis series, anode paste workshop, foundry, technical and general services), residential complexes, installations and railways, and working capital; it amounts to 8 billion francs CFA (16 billion French Franc at the time) (Laparra, 2002; Lasserre, 1958).

4 Conversions based on current US$.

5 Two ships 8000t self-unloading aluminum carriers, the Paul Héroult and the Sainte-Claire Deville shuttled between Conakry (Republic of Guinea) and Douala (Cameroon) (Laparra, 2002). At full capacity, the factory had a workforce of 550 people, with a European management of 90 expatriates (Laparra, 2002), a dominance entrenched from the late 1950s through the 1960s.

materials. Aluminum production relies heavily on bauxite, the primary ore from which aluminum is extracted, and many colonies were rich in bauxite deposits. Thus, colonial powers such as France saw the construction of aluminum processing plants in their colonies as a way of reducing the logistical costs of transporting raw bauxite over long distances to the metropoles. Processing bauxite into aluminum in the colonies allowed them to reduce significantly material transported, which was particularly advantageous in a period when shipping costs were high. Second, it was economically convenient to establish such plants in the colonies due to access to cheap raw materials and exploitable labor. Table 2.2 provides estimates related to the core costs of producing a ton of aluminum in Ghana in the 1960s.

Profitability largely depended on the costs of alumina and power, with transportation playing a key role (Manu, 2003). The distance between the bauxite mine, alumina plant, and smelter significantly influenced costs, and given that bauxite input weighed 8 to 10 times more than the final aluminum product, concentrating all three production stages geographically offered substantial economies (Manu, 2003). Also, it is no secret that colonial economies were structured to exploit indigenous labor, usually at low or no cost (Baxter, 2008; Cooper, 1996). Building an aluminum processing plant in Cameroon enabled colonial France to take advantage of a labor force subjected to low wages, harsh working conditions, and discriminatory laws. This kept the operational costs of such plants low, maximizing profits for the capitalists often based in the metropoles or operating through subsidiaries (such as ALUCAM)

Table 2.2 Core costs of producing a ton of aluminum in Ghana in 1960s

Components	Cost (in US$)
20,000 Kwh of electricity @ 3 mills/Kwh	$58.80
2 tons of alumina @ $70/ton	$140.00
14 cwt. of pitch-coal @ $50.40/ton	$36.40
160 lbs. of electrolyte @ $350/ton	$25.20
30 hours of labor	$30.80
Operating and Maintenance Supplies	$19.60
	Total $310.80

Source (Birmingham et al., 1966: 401)

controlled by the same capitalists. Third, energy availability in colonies. The aluminum production process is energy-intensive, especially during the smelting phase. Several African colonies possessed relatively untapped energy resources, such as hydroelectric power. Therefore, as indicated earlier, colonial powers invested in energy infrastructure in specific regions in their colonies not primarily for the development of those local communities but to support industrial demands, including aluminum production.

The building of hydroelectric dams and aluminum plants in the colonies fostered dependencies between the colonies and the metropole. That a significant portion of electricity generated in Cameroon was directed toward the operation of colonial firms or the export of goods to the metropole, left little power available for the development of the local economies. This structural imbalance was difficult to fix in the postcolony, hence was simply reproduced. From this angle, this chapter does more than chart a genealogy of solar power capitalism; it reinterprets Africa's energy history itself. By linking early grid exclusions, hydro-industrial complexes, and contemporary solar initiatives, I show that energy development on the continent has consistently functioned as a mode of political rule and capital accumulation. This repositioning situates electrification in Africa as central to the global history of infrastructure and empire.

Additionally, once the energy infrastructure was built, postcolonies such as Cameroon became economically tied to the industries that justified its existence. This dependency persisted as the infrastructure's operation continued to favor global industrial supply chains over local development.[6] Also, the hydroelectric dams such as that in Edea was built with expertise from colonial France, expertise which the newly independent countries did not have. So, most African countries remained dependent on foreign firms for energy management and development. The afterlives of these infrastructures stretch further still. Interviews from Ghana's bauxite mining communities and Accra's Agbogbloshie scrapyard reveal how energy projects leave toxic residues that shape

[6] As Obeng-Odoom (2016) argues, colonial infrastructures entrenched logics of accumulation that contemporary urban economies still mirror.

livelihoods long after initial construction. These sites prefigure the extractive and waste frontiers of solar energy capitalism, which I explore in Chapter 5 and return to in Chapter 6 through women's insertion under patriarchal governance. By the late 1950s, as independence movements gathered momentum, these infrastructures were already locking African economies into extractive circuits designed in the colonial era. As I show in Chapter 5, the afterlives of solar devices reveal that the promise of clean energy is inseparable from extractive and toxic circuits that persist long after solar panels have been installed.

The colonial cases make clear that infrastructure is never merely technical; once hydroelectric dams, transmission lines, and tariff schedules are laid down, they canalize flows of capital and legitimacy long after the colonial power's departure. These built environments created what scholars call path dependency (Kay, 2005; Poteete, 2009; Young, 2012): they locked post-independence governments into servicing the very corridors and concessionary contracts that had privileged foreign industry in the first place. It is across this axis of material persistence that I now reveal how familiar hierarchies were repackaged rather than dismantled. This matters because it shows how grids built to serve colonial enclaves locked in hierarchies that contemporary solar electrification projects continue to inherit. In this respect, the chapter contributes not only to debates on solar power but also to African historiography more broadly. By tracing how colonial grid design logics resurface in the architecture of contemporary solar finance, I argue that Africa's energy history must be understood as a managed history of engineered scarcity, enclosure, and selective inclusion—one that recasts familiar narratives of underdevelopment through the lens of deliberate infrastructural design.

In sum, the colonial grid inscribed a triad of governing logics: enclave privilege, risk externalization, and concessional underwriting that connected extraction to exclusion. Each logic can be read as a distinct expression of insertion and depletion: insertion through selective incorporation into industrial circuits, and depletion through the exhaustion of people and ecologies that sustained them. These design habits did not disappear with decolonization; they merely changed their institutional attire, resurfacing in contemporary solar energy frontiers.

2.3 Reworking Empire

In the early 1960s, soon after Ghana's independence, the Akosombo project revealed how new states inherited and extended these imbalances. As indicated earlier, the fact that the newly independent states lacked financial and technical capacity fostered dependency that existed during the colonial era. This dependency was brought to the fore during the Akosombo Dam construction in Ghana. The country's first president, a nationalist and pan-Africanist, Kwame Nkrumah, criticized imperialist governments for their persistent domination of Africa, manifested through capital monopolization, financial exploitation of dependent colonies, and the resulting unequal development (Nwoko, 2006), but turned to the same capitalists' network for assistance in the construction of the Akosombo dam. As part of the Volta River Hydroelectric Power Project, President Nkrumah signed an agreement that led to the establishment in September 1959 of the Volta Aluminum Company Ltd. (VALCO), owned by two US-based multinational companies—Kaiser Aluminum and Chemical Corporation (90%) and Reynolds Metal Company (10%) (Manu, 2003). This agreement led to the largest foreign direct investment (US$196 million) in Ghana in the early 1960s (Manu, 2003). The Volta project was financed by US$98 million in equity from the Ghanaian government and loans from the International Bank for Reconstruction and Development (US$47 million at 5.75% interest), USAID (US$27 million at 3.5% interest), and US Exim Bank (US$10 million at 5.75% interest) (Birmingham et al., 1966).[7]

To grasp the depth of Ghana's concessions, it is useful to examine the specific fiscal and tariff terms granted to VALCO. As per the agreement between the government and VALCO:

a. imports by VALCO for the construction of the smelter and its operation were to be duty free for the first thirty years;
b. no restrictions, control, or taxation of aluminum exports;

[7] The loans were to be repaid over 25 years for the International Bank for Reconstruction and Development and US Exim Bank, and 30 years for USAID (Birmingham et al., 1966).

c. VALCO was granted pioneer company relief which exempted it from any taxation on all its income for at least five years, extending beyond that period to a maximum of ten years to enable
VALCO to make sufficient net earnings to recover,
   i. an amount equal to the total paid-up share capital of the company, and
   ii. an amount equal to 50% of the total capital cost incurred in constructing the smelter by the second anniversary of the Permanent Delivery Date;
d. from the years when income tax did become payable (i.e., 1978) until 1997, the rate at which tax was to be assessed was that applicable to companies on January 2, 1961, which was then 40% of retained profits, plus an additional 2.5% of profits transferred out of Ghana. With a few minor exceptions no other taxes were to be levied on VALCO's operations during this period (Manu, 2003: 20).

The electricity price was initially set at 2.625 mills/kWh, increased to 2.75 mills/kWh in 1972, 3.25 mills/kWh (backdated to 1973) in 1977, and 5 mills/kWh in 1979 (Manu, 2003). The initial rate of 2.625 mills/kWh was significantly lower than the 4 to 4.5 mills/kWh typically charged for electricity used in aluminum smelting up to 1970 (Manu, 2003). Likewise, the mid-1970s rate of 3.25 mills/kWh was far below the 12 mills/kWh Kaiser paid in 1976 for coal-powered electricity at its Ravenswood, West Virginia, US smelter (Manu, 2003).[8]

In addition to the loss in revenue due to the tax concessions, Ghana was operating the hydroelectric dam at a loss. In 1976, VALCO got 2645 million kWh at 3.25 mills/kWh, generating GHS9.9 million [US$ 799,613] in revenue but at a cost of GHS10 million [US$ 807,904] for Ghana (Hart, 1980). As VALCO was using about 65% of electricity generated from the hydroelectric dam, even the World Bank, a champion for creating a conducive environment for capitalists, admitted that

[8] Electricity rates charged by the Bonneville Power Authority, United States which relied exclusively on hydroelectricity were as follows: December 1979 (5 mills/Kwh), July 1, 1981 (9.2 mills/Kwh), October 1, 1981 (17.3 mills/Kwh), October 1, 1982 (25.9 mills/Kwh) (OECD, 1983: 33).

VALCO would significantly hinder the growth of non-smelter electricity demand (World Bank, 1969). Put differently, the hydroelectric dam was not working toward improving the wellbeing of Ghanaians as well as the country's economic development. Because VALCO received electricity at an exceptionally low rate, the Ghanaian consumer was left to bear the burden of paying significantly higher rates (Manu, 2003).

Since VALCO consumed over 60% of Ghana's power while remaining disconnected from domestic industry, it foreclosed diversification into iron, textiles, or paper manufacturing (Manu, 2003). Studies on the iron ore, manganese, textile, pulp, and paper industries indicated that about 600 MW must be added to Ghana's power capacity if the country were to establish those industries (ibid.). Hence, it is important to historicize Ghana's current economic development, electricity deficits, and subsequent embrace of solar electrification. The same way that it is important to look at hydroelectrification from a critical perspective, it is vital, as I show in different chapters, to look at solar electrification from a critical perspective (see Chapters 4–6). The uneven development and structures which foster dependency, manifested during the colonial era and continued in postcolonial Ghana. What this case demonstrates is how postcolonial infrastructure, far from escaping colonial logics, often deepened dependency. These dynamics reverberate in contemporary solar energy projects, where concessional terms and enclave designs continue to privilege external actors. This matters because the Ghanaian case illustrates how postcolonial concessions like VALCO channeled limited hydro power into foreign-dominated industry at the expense of national development, a pattern echoed in later solar energy deals.

This logic extends into Ghana's recent embrace of utility-scale solar plants. Interviews from Kaleo and Lawra reveal how land allocation for the solar plant has negatively affected the livelihoods of some community members. These projects replicate the colonial pattern of enclave privilege: electricity exported into the national grid, while nearby communities remain burdened with unresolved land claims and little direct benefit. Just as households today weigh whether to pay for solar credits or purchase flour, Ghanaian citizens then were effectively subsidizing VALCO through higher tariffs. These microeconomic sacrifices point to the broader structural artifact: infrastructures that externalized risk and cost downward while consolidating profits upward. Chapter 5 returns to this tension, showing how utility-scale solar plants sit uneasily alongside extractive frontiers, producing new landscapes of dispossession and waste. Here again, the logic of concessional underwriting and managerial control emerges clearly, illustrating how financial structures were engineered to secure investor returns rather than expand universal energy access. This logic of concessional terms and enclave privilege was not unique to Ghana.

Unlike former British colonies in Africa such as Ghana, Kenya, and Malawi, the economic dependency in former French colonies such as Cameroon was more entrenched due to their political underpinnings. When Cameroon gained independence from France in 1960, it signed a monetary agreement that made it part of the Franc monetary Zone.[9] Membership of the Franc Zone during that period entailed the respect of four fundamental principles: unlimited convertibility guarantee from the French treasury, fixed parities, free transferability, and centralization of foreign exchange reserves (Kofele-Kale, 1981). The convertibility guarantee required France to ensure the convertibility of currency issued by banks, with no limits by the French treasury and in exchange, Cameroon (like other countries in the Franc Zone) was to keep 80% of its foreign exchange reserves in its Operations Account in the French treasury (Kofele-Kale, 1981).[10] Holding 80% of foreign holdings in the French treasury deprived Cameroon of the financial resources it would have used to fuel its economic development, hence deepening uneven development between the country and the global North. This monetary agreement was one of several agreements signed between France and its former colonies. On November 13, 1960, the Ahidjo government of Cameroon signed ten separate agreements with France.[11] As Corbett (1972) points out, no

[9] The Franc Zone included 14 sub-Saharan countries, the Comoros and France.

[10] It is worth noting that the 80% came because of a revision of the agreement. The original 1960 accord compelled Cameroon like other member states to hold all their reserves in French francs on deposit in the Operations Account with the French treasury (Kofele-Kale, 1981).

[11] These included a cooperative agreement on monetary, economic, and financial matters; a convention regulating relations between the Cameroonian and French treasuries; a diplomatic convention; a general technical cooperation agreement on matters of personnel; an agreement on military technical assistance; a cooperation agreement on matters of civil aviation, aerial navigation, air bases, and meteorology; a cultural convention; a convention on the role and status

one area of mutual relations was overlooked in the series of agreements France signed with each of the 14 countries when they gained 'independence.' These agreements were aimed at ensuring a continuation of French economic and political control of these countries.

A parallel story unfolded in East Africa, where dependency took new institutional forms. In Tanzania, where its president, Julius Nyerere who espoused the 'African socialism' ideology with emphasis on rural electrification could not escape the dependency structures, since the foundation had already been laid prior to the country gaining independence. Even after independence, the British government financed electrification projects in the country. After the government gained political control over the Tanzania Electric Supply Company (TANESCO) in 1964, the newly established African board appointed the British F.S Batty to head the company, and higher management of the company was made up of 63 expatriates (van der Straeten, 2023). This was unsurprising due to the limited number of qualified Tanzanians to fill technical and managerial positions.

As a further show of Tanzania's dependency on the global financial capitalist network, the World Bank made maintaining the general management of TANESCO as a condition for financing the Great Ruaha Power Project (van der Straeten, 2023). Simply put, the World Bank was preserving British interest in postcolonial Tanzania. Sweden gained a foothold in the country through development assistance and subsequently entered the growing hydroelectric market through the provision of 'Swedish technology' (Öhman, 2007: 160). Of course, Sweden was not a colonial power, but it saw Tanzania's hydroelectric sector as a fertile market for Swedish technology—or to put bluntly, Swedish capitalists. Because Tanzania depended largely on external actors for resources, it had to put aside its Wami project, a multipurpose irrigation and hydropower project, in favor of the Great Ruaha hydropower project recommended by Sweden and the World Bank (Öhman, 2007). Hence, in December 1970, it was agreed that US$43 million in financial assistance from Sweden and the World Bank will be provided for the construction of the

---

of the French military mission in Cameroon; a consular convention; and a legal convention (Kofele-Kale, 1981).

Great Ruaha hydroelectric dam, regarded as the 'largest electric power project ever to be undertaken' in the country (Öhman, 2007: 206). Sweden provided a loan on International Development Assistance (IDA) terms, to be repaid over 50 years, a 10-year grace period, and a 0.75% service charge with no interest, and the Tanzanian government would subsequently relend the loan to TANESCO for 25 years, with 5 years' grace and a 7.2% interest rate, hence leaving surplus funds within the country (Öhman, 2007).

While this represented a highly favorable credit for Tanzania, it ensured Sweden's influence over project decisions. This influence was brought to the fore when a Swedish consortium was awarded the largest contract for the construction of the power plant, reservoir, and access roads (Öhman, 2007). On its part, the World Bank provided a loan directly to TANESCO, guaranteed by the Tanzanian government, with a 25-year term, 5 years' grace, and 7.25% interest (ibid.). Even highly favorable loan terms preserved foreign managerial authority, transforming concessional finance into a subtle instrument of control. This prefigures how concessional underwriting in solar finance today binds households into repayment regimes that externalize costs downward.

As I have shown so far, various actors in the global capitalist network were deeply involved in the energy sector of newly independent African countries. Their involvement continued several decades after independence culminating in the rise of what I term solar power capitalism in contemporary Africa. This reliance on external loans and technical management exemplifies the third design logic: concessional underwriting that cements managerial control. Even favorable loan terms preserved the authority of foreign experts and financiers, ensuring that African states remained dependent rather than autonomous in their energy planning. Chapter 6 demonstrates how concessional finance translates into household-level insertion of women's credit and labor. In this way, concessional underwriting is not only a financial practice but a mode of insertion—enmeshing states and, eventually, households into externally governed repayment regimes whose hidden cost is depletion of reproductive capacities.

The postcolonial era therefore reproduced the same governing forms first inscribed under empire. Concessional underwriting, ostensibly benevolent, functioned as a financial hinge through which foreign control was re-secured. Together with enclave privilege and the displacement of ecological and social risk, these design logics constitute how insertion and depletion became the very architecture of African electrification: incorporation of states and households into debt-driven energy circuits, and the quiet erosion of resources and autonomy that kept those circuits stable.

2.4 Concluding Remarks

In examining the colonial and postcolonial genealogies of electrification, I have argued that today's solar energy frontier represents continuation rather than rupture. The enduring structures of enclave privilege, risk externalization, and concessional underwriting persist as the operative expressions of insertion and depletion that bind African households and states to global circuits of capital.

Having established this historical foundation, I now shift to the geopolitical present, illustrating how multilateral institutions, states, and corporations retool these legacies through market-making and policy design to reproduce Africa's structural dependence in the era of green energy capitalism.

References

Adebayo, A. D. (2024). Electricity, agency and class in Lagos Colony, c. 1860s–1914. *Past & Present, 262*(1), 168–206.

Adu-Gyamfi, S., Amakye-Boateng, K., Awuah, D., Oware, R., & Quansah, S. (2020). An evolutionary study of production of electricity in Ghana (1900–1960). *History of Science and Technology, 10*(1 (16)), 10–33.

Baker, L. (2023). New frontiers of electricity capital: Energy access in sub-Saharan Africa. *New Political Economy, 28*(2), 206–222.

Baxter, J. (2008). *Dust from our eyes: An unblinkered look at Africa*. Wolsak and Wynn.

Baxter, R. M. (1977). Environmental effects of dams and impoundments. *Annual Review of Ecology and Systematics*, 255–283.

Birmingham, W., Neustadt, I., & Omaboe E. N. (1966). *The economy of Ghana*. Northwestern University Press.

Byatt, H. A. (1920). From H. A. Byatt, Administrator, Dar Es-Salaam to Principal Secretary of State for the Colon[y], London, 9 June 1920. *Tanganyika Territory No. 224*. Public Record Office, Kew. PRO 161/1049.

Cecelski, E., Ounalli, A., & Aissa, M. J. (2005). *Rural electrification in Tunisia: National commitment, efficient implementation and sound finances* (ESMAP Formal Report 307/05). Joint UNDP/World Bank Energy Sector Management Assistance Program, World Bank, Washington, DC, August.

Cooper, F. (1996). *Decolonization and African society: The labor question in French and British Africa* (No. 89). Cambridge University Press.

Corbett, E. (1972). *The French presence in Black Africa*. Black Orpheus Press Inc.

DPGE [Development Plan General Electricity]. (1954, October 28). RG. 7/1/1627. Public Records and Archives Administration Department, Ghana.

Folkers, A. (2024). Risking carbon capital: Reporting infrastructures and the making of financial climate risks. *Economy and Society, 53*(3), 504–526.

Hård, M. (2023). Accessing electricity in East Africa: Dar es Salaam dwellers pursue power. In *Microhistories of Technology* (pp. 101–128). Palgrave studies in the history of science and technology. Springer.

Hart, D. (1980). *The Volta river project: A case study in politics and technology*. Cambridge University Press.

Hausman, W. J., Hertner, P., & Wilkins, M. (2008). *Global electrification: Multinational enterprise and international finance in the history of light and power, 1878–2007*. Cambridge University Press.

Jenkins, K., McCauley, D., Heffron, R., Stephan, H., & Rehner, R. (2016). Energy justice: A conceptual review. *Energy Research & Social Science, 11*, 174–182.

Kay, A. (2005). A critique of the use of path dependency in policy studies. *Public Administration, 83*(3), 553–571.

Kofele-Kale, N. (1981). Cameroon and its foreign relations. *African Affairs, 80*(319), 196–217.

Laparra, M. (2002). Enelcam–Alucam: L'énergie hydroélectrique du Cameroun à la rencontre de l'aluminium. *Outre-Mers, 89*(334–335), 177–200.

Lasserre, G. (1958). L'équipement hydro-électrique de l'Afrique tropicale française. *Cahiers D'outre-Mer, 11*(43), 232–248.

Manu, F. A. (2003). Negotiating with foreign investors: Lessons from the Ghana–VALCO experience. *Journal of African Business, 4*(1), 5–35.

Marwah, H. (2017). Electricity access inequality in sub-Saharan Africa, 1950–2000. *African Economic History, 45*(2), 113–144.

McCauley, D. (2018). *Energy justice: Re-balancing the trilemma of security, poverty and climate change*. Palgrave Macmillan.

McDonald, D. A. (2012). *Electric capitalism: Recolonising Africa on the power grid*. Routledge.

Mchome, E. L. (2022). *Blackout Blues: A socio-cultural history of vulnerable electricity networks and resilient users in Dar es Salaam, 1920–2020*. Technische Universität Darmstadt.

Nwoko, M. (2006). *Basic world political theories* (F. O. C. Njoku, Ed.). Snaap.

Obeng-Odoom, F. (2016). *Reconstructing urban economics: Towards a political economy of the built environment*. Zed Book.

Obeng, L. E. (1977). Environmental impacts of four African impoundments. *ASCE, 7*(7), 29.

Organisation for Economic Co-operation and Development [OECD]. (1983). *Aluminium industry: Energy aspects of structural change*. OECD.

Öhman, M. B. (2007). *Taming exotic beauties: Swedish hydro power constructions in Tanzania in the era of development assistance, 1960s–1990s* (Doctoral dissertation). KTH.

Ojong, N. (2022). Energy justice and off-grid solar electrification in Africa: Trends, narratives and contestations. *Off-grid solar electrification in Africa: A critical perspective* (pp. 1–21). Palgrave Macmillan.

Owens, S., & Driffill, L. (2008). How to change attitudes and behaviours in the context of energy. *Energy Policy, 36*(12), 4412–4418.

Parsons, T. (2019). *Urban development and the East African railways*. Oxford University Press.

Pinzur, D. (2021). Infrastructural power: Discretion and the dynamics of infrastructure in action. *Journal of Cultural Economy, 14*(6), 644–661.

Poteete, A. R. (2009). Is development path dependent or political? A reinterpretation of mineral-dependent development in Botswana. *The Journal of Development Studies, 45*(4), 544–571.

RCEER. (2005). *Guide to Electric Power in Ghana*. Resource Center for Energy Economics and Regulation, Institute of Statistical, Social and Economic Research, University of Ghana, Legon.

Roitman, J. (2025). Financial platforms beyond the North–South divide. In *The Cambridge global handbook of financial infrastructure* (pp. 49–63).

Showers, K. B. (2011). Electrifying Africa: An environmental history with policy implications. *Geografiska Annaler: Series B, Human Geography, 93*(3), 193–221.

Simpson, E. S. (1969). Electricity production in Nigeria. *Economic Geography, 45*(3), 239–257.

Sovacool, B. K., & Dworkin, M. H. (2015). Energy justice: Conceptual insights and practical applications. *Applied Energy, 142*, 435–444.

van der Straeten, J. (2021). Measuring progress in megawatts: Colonialism, development, and the "unseeing" electricity grid in East Africa. *Centaurus, 63*(4), 651–674.

van der Straeten, J. (2023). *Transmitting development: Global networks and local grids in the electrification of East Africa, 1906–1970* (Doctoral dissertation). Technische Universität Darmstadt.

Walker, G., & Day, R. (2012). Fuel poverty as injustice: Integrating distribution, recognition and procedure in the struggle for affordable warmth. *Energy Policy, 49*, 69–75.

World Bank. (1969). *Preliminary appraisal of the Volta River Project*. World Bank.

Worth, D. (1998, September 4–5). *Lighting the darkness: The gas supply industry in nineteenth century Cape Town, and its role in the formation of an industrial landscape.* Paper presented at the Association for Industrial Archaeology Annual Conference, North Abott, Devon.

Yenneti, K., & Day, R. (2015). Procedural (in)justice in the implementation of solar energy: The case of Charanaka solar park, Gujarat, India. *Energy Policy, 86*, 664–673.

Young, C. (2012). *The postcolonial state in Africa: Fifty years of independence, 1960–2010*. University of Wisconsin Press.

Zébazé Togouet, S. H., Nyamsi Tchatcho, N., Tharme, R. E., Piscart, C., & Wantzen, K. M. (2023). The Sanaga River, an example of biophysical and socio-cultural integration in Cameroon, Central Africa. In *River culture: Life as a dance to the rhythm of the waters* (p. 145).

3

Engineering Solar Energy Markets

3.1 Introduction

Solar energy's evolution from a fringe concept to a global economic force has been anything but linear. Far from a purely technological or environmental narrative, I show that solar power's ascent has been deeply entangled with global capitalism, strategic market-making, and the political imperatives of both the global North and South. Key actors, including major oil companies, national governments, multilateral institutions, non-Western powers such as China, philanthropic organizations, and celebrity entrepreneurs, propelled photovoltaics from scientific curiosity to profit-driven enterprise. What emerges is not a monolithic success narrative but a contested arena of divergent interests, uneven gains, and enduring tensions.

In revealing these forces, I argue that the global solar energy boom operates through recurrent moments of insertion and depletion: the incorporation of new territories, technologies, and populations into green value chains, and their simultaneous exhaustion as the cost of sustaining those chains. The question, then, is whether Africa's solar power expansion generates structural transformation or merely reconfigures old dependencies under the guise of sustainability.

N. Ojong, *Solar Power Capitalism*, Energy, Climate and the Environment,
https://doi.org/10.1007/978-3-032-13707-4_3

To develop this argument, I draw on Polanyi's (1944) insights into market engineering alongside dependency and world-systems theories, which interrogate the structural subordination of peripheral economies in the international division of labor (Amin, 1976; Frank, 1967; Wallerstein, 1974, 2004). Through this lens, solar energy markets appear not as spontaneous outcomes of innovation but as engineered circuits of insertion, in which state policies, donor agendas, and financial instruments integrate African economies into global flows of concessional capital. But this integration produces depletion of autonomy, resources, and policy space, as market-making logics lock peripheral actors in subordinate roles within the architecture of global capitalism. From this angle, solar power capitalism does not simply diffuse; it is built, engineered through mechanisms that simultaneously empower and constrain.

The argument unfolds through four interlocking dynamics; each explored in the sections that follow. First, I show how oil shocks and financial crises redirected surplus petro-capital into photovoltaics, an early form of insertion that transformed a marginal technology into a financial asset class. Second, I examine the policy scaffolding that institutionalized these markets, including tax credits, procurement schemes, and later Results-Based Financing, which embedded insertion within technocratic regimes of measurement and performance. Third, I explore the moral-market hybrids of philanthrocapitalism and celebrity electrification that aligned profit with humanitarian appeal, inserting African households and entrepreneurs into circuits of green finance while depleting the very solidarities such narratives claim to empower. Fourth, I analyze African statecraft, showing how governments appropriated external agendas for domestic ends even as structural constraints persisted, an instance of bounded insertion where agency operates within the limits set by global accumulation.

These dynamics reveal how solar power capitalism thrives on the dialectic of insertion and depletion, simultaneously opening space for participation and deepening dependency.

3.2 Redirecting Surplus Capital

> I'd put my money on the sun and solar energy. What a source of power! I hope we don't have to wait until oil and coal run out before we tackle that.
> —Thomas Edison, in conversation with Henry Ford and Harvey Firestone, 1931[1]

How did the 1970s oil crises redirect surplus petro-capital into photovoltaics and open the door to green branding? Before the oil shocks renewed global interest in alternative energy sources, solar energy had already debuted commercially.[2] However, high production costs rendered early ventures unprofitable, leading to their decline (Jones & Bouamane, 2012). The point emphasized here is the way crisis reallocates capital into new asset classes that later get exported to Africa as policy orthodoxy.

In October 1973, amid the Yom Kippur War, the Organization of the Petroleum Exporting Countries' (OPEC) production cuts quadrupled oil prices within months (Issawi, 1978), exposing the fragility of Western energy systems and catalyzing a search for alternatives.[3] From an insertion-depletion lens, these crises did not simply redirect capital; through state-led energy transitions and new technological regimes, they

---

[1] Thomas Edison, quoted in James Newton (1989: 31), *Uncommon Friends: Life with Thomas Edison, Henry Ford, Harvey Firestone, Alexis Carrel, and Charles Lindbergh*, Boston: Mariner.

[2] During the 1890s, several solar water heater companies emerged in California, a region rich in sunlight but lacking coal (Jones & Bouamane, 2012). In 1891, Clarence Kemp introduced the Climax solar water heater, which used a glass-covered pine box to trap solar heat in water tanks, and by 1900, 1600 units had been sold in Southern California (Madrigal, 2011). In 1908, William J. Bailey, a former Carnegie Steel Company engineer, developed a more efficient solar thermal collector, which featured a black metal sheet with attached pipes inside a glass-covered box, allowing water to heat faster by reducing the volume exposed to sunlight (Jones & Bouamane, 2012).

[3] On October 16, 1973, oil ministers from Iran, Kuwait, Iraq, Saudi Arabia, Qatar, and the UAE raised the price of Saudi light crude from \$3.01 to \$5.12 per barrel, increasing government revenue from \$1.77 to \$3.05, and the following day, Arab ministers proposed output reductions and embargoes on 'unfriendly states' (Issawi, 1978). On December 22, 1973, six Persian Gulf nations raised prices again, setting Saudi light crude at \$11.65 per barrel; this time, Iran led the initiative, increasing the government's share to over \$10 by the end of 1974 (Issawi, 1978).

restructured dependency, inserting peripheral economies into new cycles of technological subordination. The end of cheap oil made energy security a national concern. France and Sweden turned to nuclear programs (Friedman, 2008), while both major parties in the United States backed solar power (Scavo, 2015). The second oil shock of 1979 and the Iranian Revolution deepened this momentum (Vietor, 1984).

In the United States, major oil companies began investing in photovoltaic firms to diversify portfolios and absorb surplus petro-dollars.[4] The solar market's commercial viability enabled First Solar to triple revenues between 2004 and 2006 and raise over $1 billion through successive initial public offerings (IPOs), reaching a US$15 billion valuation by 2008 (First Solar, 2009; Sun, 2013).[5] European capitalists also entered the field, but by the early 2000s semi-peripheral actors were leveraging cost differentials and industrial policy to challenge Western dominance.[6]

American and European capital initially defined the solar industry's trajectory, but the sector's commercial viability soon invited new entrants beyond the transatlantic core. In 2000, the Australian firm, Pacific

[4] In the United States, Solar Power Corporation, a subsidiary of Exxon, was launched in 1973 (Jones & Bouamane, 2012). In 1974 Mobil entered a joint venture with Tyco to develop and sell silicon solar cells (Perlin, 2002). In mid-1977, Solar Technology International was acquired by Atlantic Richfield Oil Company (ARCO), becoming ARCO Solar (Quinn, 1985). Another oil company, Amoco, invested US$7 million in Solarex (Jones & Bouamane, 2012; Sun, 2013). By the early 1980s, the United States held 80% of the global solar market, with domestic companies making up 85% of sales, and the country exported 45% of its production, dominating the industry worldwide (Jones & Bouamane, 2012).

[5] It is worth noting that Germany accounted for 95% of First Solar's revenue in 2004, peaking at 99.5% in 2005, before declining to 95% in 2006, 90% in 2007, and 74% in 2008 (Frondel et al., 2008).

[6] In 1979, two European firms—Leroy-Somer, a French electric power generation company, and Holec, a Dutch electrical company—provided equity investment in Solarex to gain access to its expertise and a manufacturing license for photovoltaic cells, enabling them to enter the European market (Jones & Bouamane, 2012; Sun, 2013). In 1990, Siemens acquired Arco's solar business, and five years later was sold to Anglo-Dutch oil company, Shell (Jones & Bouamane, 2012; Wald, 1989). In 1994, the German-owned Ampyr Solar Europe (ASE) acquired Mobil's solar business (Jacobsson et al., 2004). In 1983, one of the main French oil companies, Compagnie Française des Pétroles (CFP), established Total Énergie in Africa to develop renewable energy, and the latter began solar module production in South Africa in 1996, and later expanded into Europe (Jones & Bouamane, 2012). Still in France, another oil company, Elf Acquitaine, and a utility company, Compagnie Générale des Eaux, established the solar firm, Photowatt, which by 1988 ranked among the top nine solar companies, contributing around 2% of global production (Jones & Bouamane, 2012; Ricaud, 2013).

Solar, launched Plug and Power systems which consisted of among other equipment, an inverter and an easy-to-mount rooftop photovoltaic system (Pacific Solar, 2001; Sun, 2013).[7] A founding director of Pacific Solar, Zhengrong Shi, left the company to start the China-based solar firm, Suntech Power, in 2001.[8] Suntech Power's 2005 New York Stock Exchange (NYSE) IPO raised US$455 million, valuing the company at US$2.5 billion (Clifford, 2015) and making Shi China's richest person (Sun, 2013). Within three years, Suntech became the world's third-largest photovoltaic manufacturer. Its ascent exemplifies semi-peripheral leverage (Arrighi, 1990; Wallerstein, 2004): firms outside the historic core exploit labor-cost advantages and state industrial policy to upend incumbents while remaining tied to core finance through NYSE listings and euro-denominated contracts. Comparable dynamics surface in Obeng-Odoom's (2024) study of oil cities, where industrial policy remains bounded by global capital hierarchies.

The oil shocks marked a turning point. They transformed solar technology from a marginal experiment into a profit-driven industry and allowed capitalists to rebrand themselves as pioneers of green innovation. What appeared as a technological breakthrough was, in Harvey's (2014) terms, a financial fix to an over-accumulation crisis. Photovoltaics functioned less as an energy hedge than as a vehicle for absorbing surplus petro-dollars—an early rehearsal of the finance-led solar energy boom that would follow decades later. Seen through the insertion-depletion lens, the oil crises were not exceptional ruptures but constitutive moments in solar power capitalism's formation: surplus capital inserted through crisis management, and social reproduction elsewhere depleted through austerity and concessional finance.

Although crisis-driven reallocation of petro-capital was the first lever, the second was deliberate policy design. In what follows, I bring to light how governments and multilateral institutions institutionalized those financial flows into durable market infrastructures.

---

[7] It partnered with Eurosolare, the Italian oil company ENI's solar subsidiary, securing it as a European distributor for Plug and Power and gaining a 25.6% equity investment (Sun, 2013).

[8] Within a year, a 10 MW crystalline silicon solar cell production line was operational, and Suntech Power quickly became profitable in 2003 due to low local labor costs, and by December 2003, an additional 15 MW line was launched (Sun, 2013).

3.3 Engineering Market Design

Which policy instruments, in practice, performed this conversion from niche technology to bankable asset class? Here, I chart the policy scaffolding across three distinct but overlapping phases—early pilots (1993–2007), the creation of market standards (2009–2016), and the later de-risking of private capital (2012 onward)—to show how policy mechanisms matured from proof-of-concept projects to full financialization. The mechanisms discussed here, such as pilots, standards, and de-risking, are not simply technocratic milestones; each is a means by which surplus capital from the global core is channeled into African energy markets, deepening the very dependency that the rhetoric of clean energy claims to overcome.

These policies are architectures of insertion: mechanisms that channel global liquidity into governable forms while externalizing the costs of depletion onto states and households.

To develop markets for solar products, capitalists required conducive market conditions. As Berndt and Boeckler (2009: 536) put it, '[m]arkets do not simply fall out of thin air but are continually produced and constructed socially with the help of actors who are interlinked in dense and extensive webs of social relations.' In other words, and following Karl Polanyi (1944), the construction of solar energy markets in the global North and South required a combination of social, economic, and political strategies and processes. Polanyi's (1944) insight that markets are 'instituted' rather than discovered dovetails with contemporary market-mechanism scholarship. In the United States, tax credits, Public Utilities Regulatory Policy Act (PURPA) contracts, and Federal Photovoltaics Utilization Program (FPUP) procurement schedules did more than subsidize demand; they codified risk and future cashflows into calculable formats that could be priced by investors. Treating legislation, quality standards, and Results-Based Financing (RBF) subsidies as mechanisms clarifies why policy design becomes a battleground: whoever develops the mechanism influences the distribution of value. But financialization alone cannot explain how these logics anchor themselves in everyday life. It is through insertion into repayment ecologies and depletion of household resources that finance translates into lived discipline. Governments have a significant role to play in the strategies and processes related to solar power market constructions by creating a conducive environment for the market to flourish. These interventions can be read as technologies of governmentality (Foucault, 1991): subtle mechanisms that render solar energy calculable, investable, and governable. They do not simply support markets; they constitute the very objects (credits, tariffs, repayment schedules) through which solar energy becomes a governable space. These instruments were not abstract. As the following cases show, concrete programs—from procurement schemes in the United States to subsidy pilots in Africa—translated the general logic of 'making markets' into everyday financial arrangements.

In the United States and other advanced economies, national and local governments played a crucial role in fostering the growth of the solar power industry through policies and financial assistance. One of the earliest and most significant interventions came in the aftermath of the 1970s oil shocks, when the US government initiated a series of legislative and financial measures to accelerate solar energy adoption. Under the Nixon and Ford administrations, Congress overwhelmingly approved key solar energy legislation, the most significant being the Solar Energy Research, Development, and Demonstration Act of 1974 (Scavo, 2015). That Act was instrumental in laying the foundation for sustained federal engagement with the solar energy industry, providing $75 million in funding to evaluate alternative energy sources, including photovoltaics (Hart, 1983). Notably, in 1978, the US Congress enacted the National Energy Act, which consisted of five acts (Charo, 1986).[9]

To stimulate private-sector engagement, the US government launched the FPUP in 1978 (Roessner, 1982). This initiative aimed to create an initial market for solar technology through government procurement, with the expectation that private demand would follow (Roessner, 1982). The FPUP facilitated large-scale purchases of photovoltaic equipment for federal facilities, injecting $12 million in its initial phase, followed by

[9] The acts include, the Public Utilities Regulatory Policy Act (PURPA), the National Energy Conservation Policy Act (NECPA), the Energy Tax Act of 1978 (ETA), the Natural Gas Policy Act of 1978 (NGPA), and the Powerplant and Industrial Fuel Use Act (Charo, 1986). The NECPA allocated further resources to commercialize solar technologies (Bradbrook, 1992; Sun, 2013).

an ambitious $98 million allocation over three years and a $1.5 billion budget over a decade (Hart, 1983). Also, the FPUP consisted of a five-year federal initiative to procure 152 MW of peak photovoltaic power for military remote power applications (Scott, 1978).[10] These measures bring to the fore the government's active role in structuring a nascent market by de-risking investments for private enterprises.

Furthermore, PURPA marked another critical shift in US energy policy by requiring utilities to purchase electricity from independent producers, including those generating power from renewable sources such as solar (Cudahy, 1995). This policy not only reduced entry barriers for solar energy firms but also introduced competition into a previously monopolized energy market. In parallel, the Energy Tax Act of 1978 provided financial incentives for solar power adoption, offering a 30% tax credit for residential consumers and a 10% credit for businesses investing in solar, wind, and geothermal energy (Jones & Bouamane, 2012). These tax incentives aimed to stimulate demand and encourage large-scale solar installations.

Dozens of US states also layered credits and exemptions in shifting combinations, showing both the experimental character of solar energy policy and the unevenness of its uptake.[11] The layering of incentives

[10] Such a large-scale order was projected to significantly lower the cost of photovoltaic cells, bringing prices down to $2–3 per watt in the first year, $1 per watt by the second year, and between $0.50 and $0.75 per watt by the fifth year (Scott, 1978).

[11] At the state level, California emerged as a leader in promoting solar energy. Under Governor Jerry Brown, the state introduced aggressive policies, including the establishment of the SolarCal office in 1978, which coordinated solar initiatives across government agencies (Jones & Bouamane, 2012). The state also implemented a 55% income tax credit for individuals and builders who installed solar energy systems, an incentive designed to drive mass adoption and achieve economies of scale (Jones & Bouamane, 2012). It is worth noting that such incentives were not unique to the state of California. As of February 1978, twenty-five US states had implemented property tax exemptions, while 14 states offered income tax credit or deduction programs—seven of which provided both types of incentives, and one state had eliminated its property tax exemption in favor of an income tax credit (Goodnight & King, 1978). Four states excluded solar purchases from sales tax, two provided refunds for sales tax already paid, three introduced accelerated amortization tax schedules for businesses installing solar equipment, and three states had established solar loan programs specifically for certain state-run institutions (Goodnight & King, 1978). The Oregon state government launched a solar loan program that allowed veterans to obtain low-interest loans of up to US$3000 for the installation of alternative energy systems (Goodnight & King, 1978), the New York State created the New York Energy Smart Loan Fund, which offered interest rate reductions of 4.5% below the market rate for loans used to finance energy-efficient upgrades and renewable energy technologies, while Florida introduced

exemplifies how policy becomes a modality of financial insertion. What appears as technocratic subsidy is, in fact, a mechanism for fixing future cash flows—governing uncertainty through household exposure and public underwriting.

The US government's solar power initiatives peaked under President Jimmy Carter, who, in response to the second oil shock of 1979, announced a $3 billion solar research program (Jones & Bouamane, 2012). This included installing a solar heating system on the White House as a symbolic gesture of commitment to renewable energy (ibid.). By the early 1980s, these efforts had transformed the United States into the global leader in solar energy, accounting for 80% of the world's solar power market and exporting nearly half of its production (Jones & Bouamane, 2012).

Beyond the United States, other Western countries also played a crucial role in supporting solar energy through policy mechanisms. Germany was an early adopter of solar energy incentives, launching a renewable energy policy in 1974 and gradually increasing subsidies for solar power development (Jones & Bouamane, 2012). By the 1990s, Germany had introduced a feed-in-tariff system that required utilities to purchase renewable energy at premium rates, providing a stable financial incentive for investors (Ćetković & Buzogány, 2016; Jones & Bouamane, 2012)—an architecture that naturalized subsidy as a form of state-led market engineering. What appeared as neutral energy policy was, in fact, the institutionalization of risk transfer: insertion through fiscal incentive, depletion through taxpayer underwriting.[12] The development of this policy framework was reinforced by state-funded pilot programs, including the '1,000 Roofs' initiative carried out between 1991 and 1995 (Lauber & Metz, 2004).

the Photovoltaics (PV) Rebate Program to overcome obstacles to PV installation and encourage its adoption by lowering the initial equipment costs (Gouchoe et al., 2002).

[12] A feed-in tariff is a policy mechanism used to promote the adoption of renewable energy, such as solar, by guaranteeing producers a fixed price for the electricity they generate and supply to the grid over a specified period, often 15–25 years. It is one of the most influential tools governments have used to accelerate investment in solar photovoltaic systems (see, for example, Cherrington et al., 2013; Rowlands, 2005).

Germany's policy template soon traveled across Europe. Spain, France, and Italy followed suit in the 2000s, introducing generous subsidies and regulatory support that fueled rapid solar industry expansion (Jones & Bouamane, 2012). Along similar lines, the UK government implemented a feed-in-tariff for small-scale renewable energy technologies with an installed capacity of up to 5 MW, which significantly boosted the solar photovoltaic sector (Smith et al., 2014).

But the scaffolds of the global North were never ends in themselves; they prepared the ground for peripheral insertion. Once stabilized in the core, these mechanisms were exported to aid-dependent economies, translating financial architecture into development orthodoxy.

In Africa, market construction leaned less on statutory compulsion and more on aid architectures, private experimentation, and hybrid procurement, producing a different, messier repertoire of mechanisms. Unlike countries in the global North, the mechanisms through which solar electrification markets were created in Africa are different and complex. This complexity arises from uneven economic development, which, as I mentioned in Chapter 1, was established during the colonial era and reproduced in the postcolony. Uneven economic development and the legacy of colonization mean that most countries on the continent often turn to external actors in the global North and other emerging economies for financial and technical resources to meet their needs. It is within this context that various actors (e.g., non-governmental organizations, multilateral organizations such as the World Bank and International Finance Corporation, international businesses, and governments) became involved in the creation and expansion of solar electrification markets in the continent. This web of actors played a critical role in enabling solar energy products development in Europe, North America, and Asia–Pacific to enter the largely untapped solar energy markets in Africa.

In the previous chapter, I showed how Sweden got access to the hydroelectric power market in Tanzania through development aid to that country. In some contexts, this strategy was used to foster solar energy markets. Specifically, international non-governmental organizations played a vital role in developing such markets. As it became clear that large-scale grid electrification could not meet the energy needs of more than 300 million people on the continent, these organizations began targeting the unserved population as recipients of (solar) energy aid.

Consider the example of SolarAid, a UK-registered international charity founded in 2006 to fight poverty and climate change through off-grid solar (OGS) electrification in the global South.[13] The charity originated from an agreement that Solar Century, then the UK's largest solar PV company, would contribute 5% of its profits to support SolarAid's work (Munro et al., 2022).[14] The partnership between SolarAid and Solar Century undoubtedly carried philanthropic overtones, but its design also hinted at a more strategic purpose. By committing 5% of its profits to SolarAid (Munro et al., 2022), Solar Century was not just donating funds; it was positioning itself as a pioneer in shaping solar energy markets in Africa.

Through SolarAid's on-the-ground networks, solar energy companies could effectively test, distribute, and promote their solar products in markets that might otherwise have been difficult to penetrate. Over time, these philanthropic endeavors developed trust among local communities, smoothing the way for solar power companies. Charitable contributions served as strategic channels for market development, aligning moral imperatives with the pursuit of profit.

SolarAid's creation of SunnyMoney in 2008 exemplified philanthrocapitalism: the blending of charitable logics with market imperatives (Bishop & Green, 2015). Established as a for-profit arm, SunnyMoney became the 'largest seller and distributor of high-quality solar lights in Africa' and has offices in Malawi and Zambia.[15] The earlier work of its parent organization, SolarAid, provided the foundation on which it built its operations.

Drawing on SolarAid's embedded experience, SunnyMoney used multiple channels to distribute its solar products, but its most effective method relied on schools and teachers (Leggett, 2019). Teachers

---

[13] As of January 2025, it reports having provided 12.5 million people with 'clean, safe, renewable light' (https://solar-aid.org/who-we-are/).

[14] Back in 2018/2019, Solar Century's profits increased by 860% in a year through investments in subsidy-free solar farms (Ambrose, 2019).

[15] https://sunnymoney.org/.

and principals received marketing materials to share with students, who in turn communicated the benefits of solar products to their parents. Families were given a two-week window to make purchases (Ekström & Eriksson, 2018). Following these campaigns, SunnyMoney staff visited schools to deliver the products. Payments followed a distinctive chain: parents gave money to their children, who passed it to teachers, who then transferred it to SunnyMoney representatives (Esper et al., 2013). Beyond the education system, SunnyMoney also partnered with local business owners in both urban centers and rural communities to sell its products. These entrepreneurs either collected stock from the company's head office or received deliveries directly (Esper et al., 2013). In effect, SunnyMoney leveraged low-cost, community-based infrastructures to expand the solar energy market.

By embedding its sales and distribution networks within educational institutions and local enterprises, SunnyMoney circumvented the logistical and trust barriers that often impede market penetration in rural Africa.[16] This strategy reduced operational costs while capitalizing on the social credibility of teachers and entrepreneurs, who served as trusted intermediaries between company and consumer. This is market-making: an insertion of moral legitimacy into accumulation. Philanthropy here functions as a soft infrastructure of finance.

In short, social infrastructures doubled as commercial infrastructure—precisely the axis on which philanthropic intent turns into market formation. Here, insertion takes a moral form: communities are woven into circuits of repayment through humanitarian rhetoric, while depletion unfolds in the unpaid time, trust, and emotional labor that sustain market uptake.

Moreover, the firm's reliance on school networks brings to the fore the importance of behavioral dynamics in fostering the solar energy markets. Parents, influenced by their children's exposure to solar power and the persuasive authority of teachers, were more likely to view solar lighting as a worthwhile investment. This strategy contrasts with conventional market-entry models, where firms typically target direct consumers through advertising and retail expansion. SunnyMoney, by contrast,

[16] On trust-building mechanisms, see Ojong (2018) and Ojong and Simba (2020).

leveraged existing social structures to generate demand and to produce early repayment and churn data that later underwrote its investability.

Such an approach aligns with broader insights from social marketing literature, which emphasize the influence of peers and trusted community figures in shaping purchasing decisions (Shah & Asghar, 2023; Wang et al., 2012), particularly in markets marked by limited awareness and affordability constraints. In this context, philanthrocapitalism did not simply 'do good while doing well' (McGoey, 2012); it actively transformed social institutions into channels of market formation, turning the provision of light into an investable object embedded within circuits of financialization. SunnyMoney's unorthodox, context-driven strategy proved effective: sales of its solar products rose from 18,000 to 51,811 within a single fiscal year (2010/2011 to 2011/2012) (Keane, 2014; Leggett, 2019).

Yet, as Thompson (2018) cautions, philanthrocapitalism extends beyond importing corporate efficiency logics into social interventions; it also enforces business rule over domains traditionally subject to public deliberation, narrowing the space for democratic decision-making. Comparable dynamics are evident in other key sectors, such as water, where international charities have been shown to commodify vital resources and reterritorialize them into commercial and financial circuits, often under the guise of empowerment (Menga et al., 2023).

In this light, SunnyMoney's school-based campaigns, while expanding solar energy access, also generated detailed repayment data. Such practices parallel the patterns observed in strategic philanthropy and financial inclusion initiatives, where ostensibly charitable ventures double as market-expansion strategies for corporate or quasi-corporate actors (see Langevin et al., 2023).

From an insertion and depletion perspective, philanthrocapitalism reveals its dual nature not as benevolent innovation but as a technology of incorporation. Moral markets insert new actors (teachers, parents, celebrities, and non-governmental organizations) into circuits of accumulation by converting compassion and legitimacy into market traction. But this very process depletes political agency: decisions once belonging to public institutions or collective deliberation are rerouted through private benevolence and branded altruism. In SolarAid's model, insertion took moral form—trust, affect, and volunteer time became conduits of value—while depletion appeared as the quiet erosion of public responsibility and the narrowing of political space to contest energy injustice. These moral markets therefore do not stand outside finance; they are its affective infrastructure, translating care into creditworthiness.

This affective infrastructure found concrete expression in SolarAid's own operations. Beneath its moral façade, the charity functioned as a channel that converted donor funds into investable, securitized cash-flows; a softer translation of the World Bank's RBF approach, but one firmly situated within the broader financialization of development. These tactics were not confined to non-governmental organizations (NGOs); they were amplified in the celebrity economy, where visibility and elite networks became additional market mechanisms.

A prominent illustration is Akon Lighting Africa, launched in 2014 by Senegalese-American musician Akon, entrepreneur Samba Bathily, and political activist Thione Niang. The initiative showed how star power can mobilize concessional finance and legitimacy while advancing a market-oriented electrification agenda. Leveraging Akon's global fame, the project attracted international financing, most notably a US$1 billion credit line from a Chinese partner, to deploy solar streetlamps and home systems across more than a dozen African countries, including Mali, Guinea, and Senegal (Akon Lighting Africa, 2016; Simmet, 2018). Framed as both humanitarian and entrepreneurial, it positioned itself within the 'base of the pyramid' discourse, claiming to empower rural communities by enabling evening commerce, improving safety, and extending study hours for children (Kranefeld, 2019).

However, as scholars of celebrity-led development observe, such ventures derive legitimacy from elite networks and the celebrity's brand, orienting accountability upward toward funders and global audiences rather than downward to local beneficiaries (Budadin et al., 2017). Akon Lighting Africa thus exemplifies how celebrity-driven solar narratives galvanize visibility and investment but risk reproducing top-down imaginaries of African electrification, where the moral authority of the celebrity substitutes for deeper engagement with structural inequalities and locally-situated energy politics.

That said, both philanthropy and celebrity, while high-profile, remained marginal relative to the scale of capital mobilized by development finance institutions (DFIs). Institutions like the World Bank and International Finance Corporation (IFC) did not simply supplement these initiatives; they standardized, scaled, and financialized solar energy markets through policy and de-risking instruments.

Specifically, the IFC, a member of the World Bank Group with a particular focus on developing the private sector in the global South, played a more significant role in catalyzing solar energy markets in Africa. The IFC and World Bank's roles have evolved significantly over the past three decades. Initially focused on development aid and electrification projects (Bernard, 2012), institutions such as the World Bank later shifted toward a market-oriented approach, catalyzing private-sector investment in decentralized solar power systems. Over three periods, the World Bank Group moved from pilots to standards and platforms, to full de-risking for private capital.

3.3.1 Period One (1993–2006): Pilot Projects

The first period focused significantly on pilot projects. During the early 1990s, the World Bank recognized that solar home system (SHS) could provide a viable alternative to grid-based rural electrification, which was often capital-intensive and slow to expand (Martinot et al., 2001).[17]

Solar power market formation also required electrification policy changes. Unsurprisingly, the World Bank, due to its financial firepower, played a vital role in this sphere in several countries. African countries in need of financial assistance, turned to the bank, which in turn asked these countries to make certain policy reforms. In the 1990s, Senegal had to implement policy reforms in the electrification sector in return for financial assistance from this Bretton Woods institution (Trompette

[17] This realization led to the approval of 12 pilot projects between 1993 and 2000 under the Global Environment Facility (GEF), aimed at enhancing SHS markets by reducing market barriers, supporting policy development, and providing subsidies to improve affordability (Martinot et al., 2001). These projects spanned multiple countries, including Kenya, Benin, Togo, and Cape Verde, and introduced several approaches such as dealer sales models, energy-service concessions, and microfinance-backed consumer credit programs (ibid).

et al., 2022). The ensuing reforms sought to enhance private investment and efficiency by implementing partial deregulation (Diouf & Miezan, 2021; Mawhood & Gross, 2014). Law No. 98-29, enacted on April 14, 1998, introduced a new legislative framework for Senegal's electricity sector, restricting the Senegalese Electricity Distribution Company's (SENELEC) authority to areas that were already electrified, most of which were connected to the national grid (Trompette et al., 2022).[18] Put bluntly, that piece of legislation paved the way for the development of solar energy market in that country.

3.3.2 Period Two (2007–2016): Standards and Market Infrastructure

The second period was centered on establishing market infrastructure and industry standards. What began as scattered pilots in the first period next consolidated into platforms: shared standards, quality assurance, and finance logics that could be scaled. A major project was Lighting Africa, launched in 2007 by the World Bank and IFC. Unlike earlier projects, which were structured around subsidies and aid, Lighting Africa emphasized market-based solutions, aiming to create a self-sustaining solar energy industry by promoting business models, improving quality assurance, and fostering consumer awareness (Baptista & Plananska, 2017; Ockwell & Byrne, 2017). This marked a significant shift from the moral economy of the 'development gift' to a more commercialized approach (Cross, 2019; Cross & Neumark, 2021).

A central component of this transformation was the creation of Lighting Global quality certification (later rebranded as VeraSol) which divided the market into branded and generic segments and made solar assets legible to investors.[19] As the OGS market began to take shape in the late 2000s, the World Bank and IFC played a pivotal role in defining quality assurance standards to protect consumers and encourage

[18] The national electricity company (SENELEC) is known in French as the Société nationale d'électricité du Sénégal.

[19] https://verasol.org/about-verasol/.

private-sector investment. The certification process formalized a bifurcation between 'branded' (quality-certified) and 'generic' (uncertified) products (Munro et al., 2022). These standards enabled investors to differentiate between product classes, thereby reinforcing confidence in the sector and attracting large-scale private capital.

During this period, multilateral financial institutions also helped lay the groundwork for financial models that would support mass adoption of SHSs (Table 3.1). The pay-as-you-go (PAYG) rent-to-own model, which allows consumers to make incremental payments for SHSs, was actively promoted by the IFC and World Bank as a key financing mechanism for rural electrification (Munro et al., 2022). Starting with small-scale pilots, PAYG financing rapidly became the dominant model for SHS distribution in Africa. Investment in PAYG enterprises grew from US$40 million in 2012–2013 to more than US$500 million between 2014 and 2016 (Lighting Global, 2016, 2018).

Furthermore, the establishment of the Global Off-Grid Lighting Association (GOGLA) in 2012, supported by the World Bank, IFC, US Department of Energy, and the UN Environment Program, marked another turning point (Harper, 2021). GOGLA functioned as an industry association that provided data, policy recommendations, and networking platforms for OGS companies (Harper, 2021; Munro et al., 2022). This initiative formalized the private sector-led approach to OGS expansion, signaling the full embrace of a market-driven energy

Table 3.1 OGS investment in the global South (excluding results-based financing)

	Grant	Equity	Debt	Total
2017	US$17 million	US$133 million	US$179 million	US$307 million
2018	US$5 million	US$123 million	US$229 million	US$357 million
2019	US$7 million	US$133 million	US$164 million	US$304 million
2020	US$29 million	US$81 million	US$206 million	US$316 million
2021	US$9 million	US$120 million	US$328 million	US$457 million
2022	US$11 million	US$395 million	US$340 million	US$746 million
2023	US$16 million	US$128 million	US$285 million	US$425 million
Total	US$94 million	US$1.2 billion	US$1.7 billion	

Notes Data sourced from ESMAP (2024)

access model. In June 2016, prominent development assistance organizations such as the now-defunct United States Agency for International Development (USAID), the Department for International Development (DFID), now known as the Foreign, Commonwealth & Development Office (FCDO), and the Shell Foundation, introduced the 'Scaling Off-grid Energy' initiative. This initiative aimed to drive expansion in Africa's off-grid energy sector and provide 'clean,' 'modern,' and affordable electricity to 20 million households (USAID, 2017). With mechanisms and data in place, the next move was obvious: de-risk the asset class and crowd in large volumes of private capital.

3.3.3 Period Three (2012 Onward): De-risking Private Capital

The third period saw an explosion in international investments, largely facilitated by the World Bank, which strategically de-risked private capital investments in stand-alone solar energy markets. Blended-finance vehicles, guarantees, and concessional loans that shifted risk off investors and onto aid-backed institutions. During this time, conventional private equity and venture capital firms based in the United States and Europe began to take an interest in Africa's expanding solar energy sector (Cross & Neumark, 2021). These investment funds, backed by institutional investors and affluent individuals, were attracted to solar energy markets due to the underperformance of other asset classes in the aftermath of the 2008 global financial crisis, coupled with the potential for growth in emerging economies (Cross & Neumark, 2021). Major investment strategies took the form of equity investments, debt financing, and risk guarantees. These de-risking instruments instantiate the argument that policy does the quiet work of converting uncertain technologies into investable assets, allowing global capital to find new frontiers for accumulation while African economies remain structurally subordinate.

Additionally, the World Bank and IFC made specific investments in various countries, thereby developing the solar power market in those countries. These investments were necessary to send a signal to investors on the World Bank Group's confidence in the OGS sector.

The World Bank's funding mechanisms included grants, RBF, and credit lines (ESMAP, 2024).[20] Interestingly, the World Bank, which for decades had pressured countries in the global South to remove subsidies on basic foodstuffs and services benefiting low-income consumers because they affect the efficient use of resources (Danaher, 1994; Sadasivam, 1997), has been behind the RBF steering wheel. In its flagship report, it notes:

> RBFs are increasingly being used to provide end-user subsidies, helping to address affordability challenges. End-user subsidies lower the cost of a product for consumers, specifically targeting those with very low incomes to help them access off-grid products [...] End-user subsidy programs typically allow customers to pay 30 - 90% less than the unsubsidized retail price of OGS products, with the average subsidy in the 60 -70% range. (ESMAP, 2024: 68)

While the immediate benefit of such subsidies is to reduce the retail cost of decentralized solar goods, the longer term benefit lies in creating financially viable OGS markets that attract private-sector players. Put differently, for the World Bank, subsidies are bad when they mainly benefit low-income consumers but are fine when they benefit profit-making firms. The stand-alone solar energy market cannot grow if there are no consumers, hence subsidies stabilize and expand the OGS market by rendering OGS goods affordable.

As shown in Fig. 3.1, the World Bank has consistently provided funding for the OGS sector on an annual basis since 2017. Between 2017 and 2024, it provided a total of US$2.4 billion in funding to the decentralized solar power sector in the global South, with most of it going to Africa. IFC provided equity investment in OGS companies such as M-Kopa, d.light, and Greenlight Planet, enabling them to scale operations across multiple African countries (Fehrenbacher, 2014;

[20] 'RBF is also referred to as pay-for-performance and performance-based funding [...] refer to the transfer of money or material goods conditional on taking a measurable action or achieving a predetermined performance target' (Eichler, 2006, cited in Oxman & Fretheim, 2009: 70–71). In the OGS sector, '[r]esults-based financing is when [solar] companies are paid based on achievement and verification of a predefined result, typically a sale' (ESMAP, 2024: 53).

Price, 2016).[21] Specifically, the World Bank provided funding for the Kenya Off-Grid Solar Access Project (KOSAP), a key initiative led by the Ministry of Energy designed to deliver electricity and clean cooking technologies to remote, sparsely populated, and historically underserved regions of the country (Ojong, 2022). Likewise, Rwanda's OGS sector has benefited from both financial and technical assistance from the World Bank. In September 2020, the World Bank's Board of Directors approved $150 million in funding (US$75 million as loan and US$75 million as grant) to 'improve access to modern energy for households, enterprises, and public institutions in Rwanda and to enhance the efficiency of electricity services' (World Bank, 2020a).[22] In a featured article related to Ghana's US$220 million GEDAP project, established to expand access to renewable energy through stand-alone solar goods and services, the World Bank noted: 'With a strong push for solar energy from the World Bank and the International Finance Corporation, life for thousands of Ghanaians is beginning to change' (World Bank, 2020b).[23] Thus, the World Bank is acknowledging that it was fostering the solar energy access market in that country. In that same document, the World Bank added:

> GEDAP has taken additional measures to ensure that the legal and regulatory framework was established for a nascent renewable energy sector. The enactment of the Renewable Energy Law supported the activities of the Bank and other donors in this sector. (World Bank, 2020b)

[21] Several nations, such as Benin, Burkina Faso, Cameroon, Cabo Verde, the Central African Republic, Côte d'Ivoire, Chad, The Gambia, Guinea-Bissau, Guinea, Liberia, Mali, Mauritania, Nigeria, Niger, Senegal, Sierra Leone, and Togo, have initiated solar projects and programs with support from the World Bank (Ojong, 2022).

[22] This financing supported the Rwanda Energy Access and Quality Improvement Project (EAQIP), which seeks to increase 'grid connections for residential, commercial, industrial, and public sector consumers' and provide grants to lower the cost of SHSs (World Bank, 2020a).

[23] As part of the Ghana Energy Development and Access Project (GEDAP), five pilot mini-grids were established to generate electricity from solar energy, serving remote communities located on islands within the Volta Lake and along the Volta River (World Bank, 2020b).

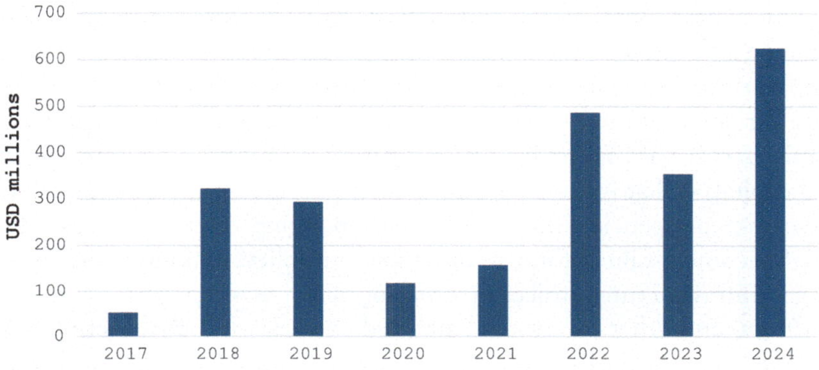

Fig. 3.1 Approved World Bank funding for off-grid goods and services (US$ millions) (Data sourced from ESMAP 2017–2024)

Here, the bank is admitting that it leveraged its financial position to push the Ghanaian government to enact solar-energy-conducive legislation. Recently, the World Bank and African Development Bank (AfDB), in collaboration with other partners, launched Mission 300—an initiative with the goal of providing 'access to electricity to an additional 300 million Africans by 2030' (World Bank, 2025). The financial assistance provided by these institutions comes with a policy conditionality, i.e., beneficiary countries are required to put in place policies to attract private-sector investment in the energy sector (see also Sguazzin, 2025). The IFC focused on the growth of the private sector by providing US$20 million loan to Daystar Power, a major provider of solar power solutions to businesses in West Africa in 2021 (Daystar Power, 2021).[24]

To facilitate the growth of solar energy access markets in Africa, another member of the World Bank Group, the Multilateral Investment Guarantee Agency (MIGA) offered guarantees to investors and lenders.[25]

[24] That loan is structured as follows: US$10 million from Canada-IFC Renewable Energy Program for Africa and US$10 million local currency loan from IFC (Daystar Power, 2021).

[25] For example, MIGA issued guarantees totaling US$43 million to cover equity and quasi-equity/shareholder loan investments by AIIF 3 Clean Energy (AIIF 3) in Bboxx's subsidiaries in Kenya, Rwanda, and Democratic Republic of Congo, including a non-shareholder loan provided by the Facility for Energy Inclusion Off-Grid Electricity Access Fund (FEI-OGEF) to Bboxx DRC in 2020 and 2021 (ESMAP, 2024). These MIGA guarantees provide protection

The World Bank, IFC, and MIGA have deployed a multifaceted strategy to create and expand solar energy markets across Africa. In other words, these multilateral development institutions played an 'escorting' role for international capital through risk mitigation. As Carroll and Jarvis (2014: 538) frame it, 'risk mitigation for capital (including financial capital) – that is the escorting of international capital by multilateral development agencies into frontier and emerging market settings – has itself become a valued form of development policy, making many (often large) infrastructure projects [*including solar power projects – author's addition*] and other investments proceed with alacrity.' By covering risks such as war, civil disturbance, and currency inconvertibility, MIGA reduces the perceived hazards for private investors, effectively widening the pool of potential financiers willing to enter these underserved countries. Such guarantees not only catalyze capital inflows but also signal international confidence in OGS energy ventures, helping to anchor longer-term market confidence. Here, depletion becomes normalized as the hidden subsidy through which solar energy markets are stabilized.

MIGA's interventions across Africa highlight how targeted risk-sharing instruments, when strategically deployed, can accelerate the expansion of decentralized solar energy markets. The IFC's equity investments in several solar power companies exemplify a deliberate approach to scale OGS energy models by empowering private-sector actors to overcome entrenched market barriers. Meanwhile, the World Bank has not only injected capital into government-led projects but also leveraged its financial clout to induce regulatory change. These strategies reveal how a blend of direct investments, technical assistance, and policy advocacy is being used to address both the supply and institutional challenges that impede the expansion of solar energy markets on the continent. Concessional flows thus rely on a double movement: the insertion of households into credit regimes, and the depletion that makes repayment appear seamless to investors.

In operational terms, escorting comprised political risk guarantees through MIGA, loans with conditionalities that rewrote electrification

---

to AIIF 3 and FEI-OGEF against non-commercial risks such as war, civil disturbance, and currency inconvertibility and transfer restrictions (ESMAP, 2024; MIGA, 2021).

policies, quality standards that sorted investable from non-investable goods, and blended-finance vehicles that shifted currency and default risk away from equity holders. To clarify these roles, Table 3.2 summarizes who did what, with what instrument, pairing each World Bank Group entity with its primary mechanism and a concrete country example.

These strategies paid off, as solar energy markets are emerging and growing across the continent, with (1) millions of solar products sold each year, (2) investments pouring into the industry, and (3) national governments 'reforming' (as the World Bank frames it) their energy policies to make room for solar energy goods and services (Fig. 3.2). In what follows, I elaborate on each of these three outcomes in turn.

Table 3.2 Escorting instruments and country examples

Institution	Instrument	Country Example
World Bank	Loans with conditionalities	Senegal: Law No. 98-29 reforms SENELEC remit
IFC	Equity/quasi-equity financing	Kenya: M-Kopa expansion
MIGA	Political risk guarantees	Democratic Republic of Congo: Guarantees for Bboxx subsidiaries
World Bank/IFC	Quality standards (Lighting Global, later VeraSol)	Pan-African: Certified vs. generic solar products

Sources ESMAP (2024), Fehrenbacher (2014), MIGA (2021), Munro et al. (2022), and Price (2016)

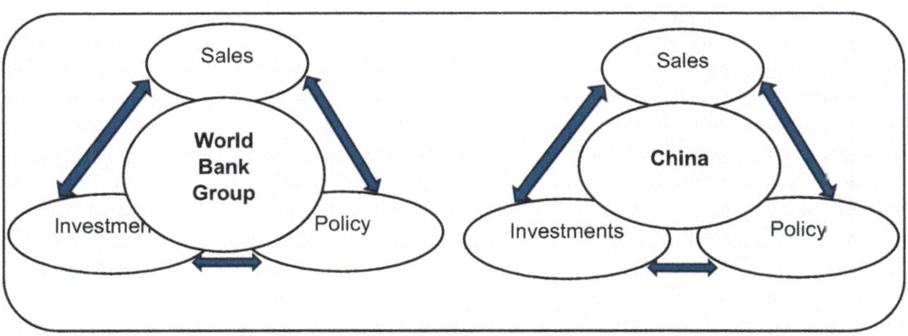

Fig. 3.2 A self-reinforcing loop among sales, investment, and policy reforms in Africa's solar energy access markets

By the early 2020s Africa's OGS market had become a multi-billion-dollar industry, driving record global sales of OGS kits. Sales expanded rapidly across both rural and peri-urban areas, reflecting the successful diffusion of solar products through a combination of donor-backed programs, private distribution networks, and consumer finance models.[26]

In parallel, the deregulation of energy sectors and rising demand for decentralized electricity led to a surge in investment in solar power sector across the global South, particularly in Africa. By the early 2020s, Africa's OGS market was attracting over a billion dollars in private investment and supplying electricity to hundreds of millions of people (Table 3.1).[27]

---

[26] Regarding sales, globally, over 50 million OGS goods were sold across 2022 and 2023, exceeding US$3.3 billion for 2019 and the market generated revenues of $3.9 billion in 2022 and $3.8 billion in 2023 (ESMAP, 2024; World Bank, 2024). Based on industry estimates, over 400 million people in Africa obtain electricity from SHSs (*The Economist*, 2024). In 2023, East Africa was the largest market for branded OGS sales, accounting for about 46% of the total global volume (ESMAP, 2024). Sales in East Africa increased from approximately 4 million units in 2021 to around 5 million in 2022, marking an 18% rise (ESMAP, 2024), and between January and June 2024, nearly three million OGS kits were purchased across that subregion (GOGLA, 2024). Between 2010 and 2018, half a million people had PAYG contracts with OGS firms in East Africa (GOGLA, 2018). In West Africa, 516,000 units of OGS kits were sold in the first half of 2024 (GOGLA, 2024). Still regarding sales, the market for solar electrification paved the way for other solar goods such as solar water pumps, refrigerators, TVs, and fans. In 2023, a total of 1.76 million fans and televisions were sold in the global South, along with 30,000 solar water pumps and refrigerators, and while sales of TVs and refrigerators have remained relatively steady, demand for fans and solar water pumps has been experiencing significant growth (ESMAP, 2024). In East Africa, 71, 127 units of solar-powered appliances (TVs, fans, solar water pumps, and refrigerators) were solar in the first half of 2024 and 191, 700 units were sold in West Africa during that same period (GOGLA, 2024). It is worth stressing that the discussion here on sales of OGS products excludes generic or nonbranded products, which have captured a significant portion of the market across Africa (Lighting Global, 2020; Samarakoon et al., 2021). The generic off-grid solar market accounts for approximately 72% of the total market share (Lighting Global, 2020). In spite this impressive growth in solar energy access markets, the World Bank acknowledges that extending existing grid remains the most cost-effective means to provide access to electricity for over 100 million people on the continent (*The Economist*, 2024).

[27] As shown in Table 3.1, between 2017 and 2023, investors poured over US$1 billion in equity investments and US$1.7 billion as debt in the OGS sector. To use the title of a recent article published by *The Economist*, 'Private firms' are driving a revolution in solar power in Africa (*The Economist*, 2024). These amounts deviate significantly from grants (US$94 million) provided during the same period, indicating strong interest from financial capitalists to make profit. Between 2017 and 2023, annual financing for the OGS sector hovered between US$300 and US$400, except for 2022. The 2022 anomaly was because of Sun King's Series D equity investment round which raised US$330 million, with participation from private

Collectively, these mechanisms reveal solar power capitalism's deeper logic: insertion through policy architecture, and depletion through social reproduction (see also Rai et al., 2014). Portfolio-level instruments that enable market growth at scale land in households as repayment schedules, remote lockouts, and data-driven scoring systems that reallocate risk downward. These dynamics unfold most starkly in everyday life, a process I examine in detail in Chapter 4. The rest of the chapter charts this shift from metrics to politics: first China's hybrid template, then African statecraft, before situating these maneuvers within longer trajectories of structural dependence.

Having shown how Western institutions standardized and de-risked markets, we can now compare this with a parallel architecture emerging from the East. China's bundled approach demonstrates a qualitatively different mode of market-making. China represents a hybrid profile: neither simply a core-like investor nor a benevolent South–South partner, but a complex actor whose bundled model unsettles simple dichotomies. Put differently, as Fig. 3.2 illustrates, alongside

---

equity investors M&G, General Atlantic, and Leapfrog (BII, 2023). Between 2010 and 2018, an estimated US$1.64 billion in equity and debt—encompassing loans, venture capital, and bonds—was directed toward what financial experts refer to as 'global energy access markets' (Burger, 2019; Wood Mackenzie, 2019). Although these investments are theoretically global, most of the funds, approximately US$1.28 billion or 75% of the total, were allocated to companies operating in sub-Saharan Africa (Cross & Neumark, 2021). Between 2017 and 2019, several major capitalists made significant investments in Africa's solar power market. In 2017, Royal Dutch Shell invested in SolarNow, a Ugandan company supplying SHSs in both Uganda and Kenya (Hanafusa & Sese, 2019), and the following year, the European Investment Bank (EIB) signed a US$25 million financing agreement with d.light, a key solar company operating across Africa (EIB, 2018). In 2019, Japan's Mitsubishi Corporation played a central role in a US$50 million funding round designed to support BBOXX, a prominent PAYG SHS provider with an extensive footprint in Africa (*Financial Post*, 2019). That same year, Azuri Technologies, a leading supplier of SHSs to off-grid households across the continent, secured a strategic investment of US$26 million, led by Marubeni Corporation, a Fortune Global 500 company (Azuri Group, 2019). In 2022, the oil giant, Shell, acquired Daystar (Shell, 2022). Recent investments provided by funders relatively new to the OGS sector include a US$10 million equity investment by Convergence Partners in Yellow Solar (2023), a US$7.9 million equity investment by At One Venture (alongside others) in Okra Solar (2023), a US$127.4 million debt investment in Sung King by CitiBank Kenya (alongside others) in 2023, a US$15 million equity investment in BBOX by the Private Infrastructure Development Group (2022), a US$75 million debt investment in d.light (over 2022 and 2023) by the Eastern and Southern African Trade and Development Bank, a US$12.5 million equity investment in Ecozen by Nuveen (2023), and a 10 billion Nigerian Naira securitized financing facility for d.light by Chapel Hill Denham in 2024 (ESMAP, 2024).

Western actors, a non-Western country advanced a different template for deploying solar power at speed.[28]

Under the Forum on China–Africa Cooperation (FOCAC), Beijing has pledged substantial resources to renewable energy, including the 2024 'Africa Solar Belt' initiative aimed at expanding utility-scale solar power projects and stand-alone solar systems across the continent (Song & Ireri, 2024). Chinese engagement is characterized by a bundled model that integrates concessional finance, engineering–procurement–construction (EPC) contracts, and technology supply, often through state-owned enterprises or private firms with state backing, thereby enabling rapid project delivery but limiting local participation in manufacturing and higher value segments of the value chain (Lema et al., 2021). China's solar cooperation is increasingly framed within its Belt and Road 'Green Development' agenda, signaling a gradual pivot from fossil fuel heavy investments to alternative energy pathways (Kiryakova et al., 2025). Financing spans commercial lenders and blended-finance structures, enabling a broad project spectrum from grid-scale plants to mini-grids (ibid.).

However, fragmented coordination between Chinese ministries, financiers, and firms can undermine host-country localization priorities, echoing critiques of 'bounded benefits' observed in earlier studies (Kiryakova et al., 2025).[29] Unlike many Western players, China's solar power investments are generally free of political conditionalities and framed within South–South cooperation rhetoric, but they reproduce

[28] China's role in Africa's solar energy access markets illustrates a mutually reinforcing nexus of investment, policy, and sales (Fig. 3.2). Concessional loans from Chinese policy banks are frequently tied to procurement from Chinese contractors and technology suppliers, embedding policy influence through financing structures that privilege Chinese firms. This arrangement sustains the global dominance of Chinese solar manufacturers, whose cost-competitive panels both enable the rapid scaling of African projects and reinforce demand for continued Chinese-backed investment.

[29] Case studies, such as the Garissa solar plant in Kenya—funded by a $135 million concessional loan from China Eximbank and built by China Jiangxi Corporation for International Economic and Technical Co-operation, in collaboration with Chinese solar company JinkoSolar Holding and Kenya's Rural Energy Authority (Kiryakova et al., 2025)—reveal that while projects create some local jobs, deliver training, and improve energy access, they remain highly dependent on imported labor, equipment, and turnkey project management from China (Bhamidipati & Hansen, 2021).

some enclave characteristics familiar from other Chinese infrastructure projects, constraining broader economic spillovers (Shen & Power, 2017). While these ventures avoid many of the displacement and environmental harms associated with fossil fuel projects, they continue to attract scrutiny over transparency and procurement practices (Kiryakova et al., 2025).

Chinese-backed renewable projects in Africa, including solar, sometimes face criticism over opaque contract negotiations and limited competitive tendering compared to multilateral or Western-funded projects, which more often follow open bidding frameworks (Kiryakova et al., 2025). These issues are linked to the prevalence of government-to-government agreements and EPC-plus-financing packages, where procurement is tied to the financier or contractor, reducing transparency and limiting independent oversight (Kiryakova et al., 2025). This hybrid profile, combining rapid deployment capacity with limited localization and governance challenges, complicates the dominant narrative that China's role in Africa's energy sector is uniformly extractive. Instead, it suggests a more variegated set of outcomes shaped not only by local agency, host-country policy frameworks, and contractual terms of engagement, but also by the degree of transparency and openness in procurement processes.

What distinguishes China's bundled model from the World Bank's 'escorting' is the locus of risk and the rhythm of project delivery. Whereas the Bank socializes risk through layered guarantees and long negotiation cycles to create investable markets, Chinese state-backed EPC firms internalize financing and construction within a single package, allowing rapid build-out but limiting host-country bargaining power over technology transfer and local manufacturing. This contrast reveals that China's 'hybrid' role is not simply a middle ground between North and South, but a qualitatively different pathway of market-making whose speed and opacity create distinct challenges for African industrial policy. If China's bundled model reshapes market-making through speed and centralized execution, the World Bank's influence is exercised more diffusely through regulatory reform and institutional restructuring.

A third area where the World Bank claims success lies in the reform of national policy frameworks to accommodate OGS electrification;

reforms that, while expanding the sector's visibility within energy strategies, also reflect the Bank's capacity to steer policy change in line with market-oriented priorities. Several African countries have updated existing energy policies.[30]

The three outcomes of the World Bank, IFC, and MIGA I just discussed are deeply intertwined in a feedback loop of market expansion and institutional support. Growing sales volumes not only boost the profit margins and market credibility of solar power firms but also attract investor interest. These higher levels of investment enable solar power companies to scale operations and refine their services, further fueling consumer demand. At the same time, as governments across Africa enact or reform policies that favor the sector, such as lowering tax and regulatory barriers, these reforms reduce costs and uncertainties for industry players, making solar power ventures even more attractive for financial capital. Also, supportive policy reforms lower market-entry barriers, thereby making OGS products and services more affordable, further fueling consumer demand (Fig. 3.2). Notably, over time, this synergy—whether in the case of China or the World Bank Group—not only transforms national energy landscapes but also reinforces their role as powerful architects of the so-called clean energy narratives shaping Africa's energy future.

But centering external architectures risks flattening domestic politics. The next section restores them, asking how African governments maneuver within, while often reproducing, the structures built around them.

[30] Kenya's Green Growth and Investment Plan, launched in 2023 explicitly recognizes the vital role of OGS electrification (Sustainable Energy for All, 2023). Rwanda's revised National Electrification Plan has a goal of connecting 48% of households through OGS electrification and micro-grids (EDCL, 2023). Uganda updated Energy Policy stresses the key role of the OGS sector in rural electrification (MEMD, 2023). Additionally, some countries have provided tax incentives, such as VAT and duty exemptions, to stimulate OGS markets by reducing the prices of OGS products. Nigeria, Sierra Leone, and Malawi have VAT or import duty exemptions in effect for OGS products (ESMAP, 2024; Kinally et al., 2023). In 2024, Cameroon introduced customs duty exemptions for essential photovoltaic components used in rooftop solar systems (Beyer, 2024), while Liberia extended an executive order suspending tariffs on OGS products (RoL, 2024).

3.4 Statecraft in Dependence

The preceding sections highlighted how crises, policy scaffolds, and external capital shaped Africa's solar energy markets. But these dynamics did not unfold in a vacuum. African governments actively mediated them, exercising agency that was real but structurally circumscribed. Dependency theorists have long argued that participation in externally financed projects seldom alters the structural position of global South economies within the global system (Amin, 1976; Cardoso & Faletto, 2024; Dos Santos, 2019). From this perspective, the embrace of hydroelectric dams or solar electrification initiatives secured short-term developmental gains but did little to transform the continent's subordinate role in the international division of labor.

However, critiques of this structuralism caution against reducing African states to passive victims of global capitalism (Cooper, 2002; Evans, 1979). A more relational analysis highlights the dialectical interplay between structure and agency: African leaders strategically appropriated external resources to pursue domestic goals, but often in ways that simultaneously reproduced class inequalities at home and dependency abroad.

For instance, electrification projects frequently privileged urban areas and priorities of political elites, often to the detriment of marginalized communities (Bhattacharyya, 2012; ESMAP, 2005; Pedersen & Poncian, 2024), exacerbating social stratification even as they expanded energy access. In Mozambique, large-scale rural electrification and grid extension projects serve as both material and symbolic means for the state to assert its presence and forge connections with rural residents (Power & Kirshner, 2019). Hence, agency is real, but its exercise often reinforces domestic hierarchies while leaving Africa's global subordination intact.

The Ghanaian case, mentioned in the previous chapter, illustrates this complexity. The Volta hydroelectric dam was not simply the imposition of the World Bank or US investors, but also a project central to Kwame Nkrumah's nationalist vision of industrialization and political consolidation. While Western capitalists reaped significant benefits, Nkrumah saw electrification as indispensable to diversifying Ghana's economy away from its dependence on cocoa exports and as a bulwark against political opposition (Brew-Hammond, 1996; Tsikata, 2006). From this standpoint, the dam project exemplified the double-edged nature of African agency: it represented an assertion of sovereignty and ambition for structural transformation, but remained embedded in financial arrangements and industrial linkages that connected Ghana to external actors.

A similar pattern emerged in Tanzania under Julius Nyerere. Hydroelectric dam construction resonated with his broader *Ujamaa* vision of socialist self-reliance, reinforcing state control over infrastructure and reducing reliance on imported goods, as captured in the following excerpt by Julius Nyerere:

> Schemes such as this one [Hale Hydroelectric Plant] are in fact the bricks and mortar evidence of the revolution which our country is deliberately and purposefully undergoing. It represents the application of science to the needs of the people. And it does this in such a way that our whole country takes further steps out of the poverty which now imprisons it. For this hydro-electric station is an example of the combination of brains, scientific knowledge, sweat and discipline which will in practice transform our nation. (Hoag, 2003: 180)

At the core of *Ujamaa* was the idea of self-reliance, both political and economic (Nyerere, 1987).[31] The electrification of Tanzania was envisioned as a means to break free from colonial-era economic structures that left the country dependent on foreign energy sources and manual labor-intensive industries (Mchome, 2022). Hydroelectric power symbolized modernity and progress, reinforcing the legitimacy of Nyerere's government (Hoag, 2003). The construction of the dam stood as a visible testament to the state's capacity to deliver large-scale infrastructure, vital for maintaining political stability and public trust (Öhman, 2007). The dam was celebrated as a national achievement, bolstering the ruling party's narrative that Tanzania was advancing toward self-sufficiency and industrialization (Öhman, 2007).

But the materialization of these projects depended heavily on foreign capital and technical expertise, particularly from Sweden and the World

---

[31] For a detailed discussion of Julius Nyerere's African socialism, see Nyerere (1962) and Nyerere (1971).

Bank. Although the dam became a symbol of modernity and political legitimacy, it also deepened Tanzania's financial exposure to external creditors and entrenched new forms of dependence (Hoag, 2003; Öhman, 2007). Hence, while agency was exercised, it was conditioned and circumscribed, illustrating the relational dynamic between global structures and national strategies. Electrification was therefore neither imposed nor sovereign but a site where external constraints and local aspirations intersected.

This relational framing is equally salient in the case of solar electrification. This pragmatism manifests along several dimensions: fiscal, political, and developmental, each revealing how African states translate structural constraint into tactical maneuvering. Through these layered adaptations, governments seek to reconcile external dependence with domestic legitimacy. African governments have not been passive recipients of OGS technologies promoted by international NGOs, donors, and multilateral development institutions. They embraced these solar goods and services for pragmatic reasons: First, African governments have been supportive of especially OGS electrification because it alleviates fiscal burdens. Specifically, most governments across the continent have positioned themselves as enablers rather than direct providers of OGS services. But this orientation cannot be understood as a sudden policy switch. It is rooted in the legacy of the 1980s structural adjustment programs imposed by the World Bank and International Monetary Fund (IMF), which rolled back the role of the state in production and service delivery and promoted market-oriented reforms.

Rwanda's trajectory highlights the political dimension of solar power pragmatism. The Kagame government pledged universal electrification as part of its Vision 2020 agenda (Ministry of Infrastructure, 2016), balancing national grid extension with OGS energy to maintain regime legitimacy. By selectively aligning donor finance with domestic developmental goals, the state both absorbed external funds and reinforced its claim to be a capable developmental authority, even as implementation depended heavily on foreign actors.

This case highlights the tactical agency that African governments exercise in shaping solar energy adoption. But dependency theorists argue that such maneuvering seldom disrupts the continent's subordinate structural position within the global economy (see Amin, 1976; Frank, 1967). What is at stake is not simply whether African governments chose to act as enablers or providers, but whether the adoption of such a development model meaningfully alters entrenched dependence. In other words, the critical question is whether these policies can generate transformative outcomes or whether they repackage the burden of provision, shifting costs and risks to individuals and private firms, without changing Africa's subordinate position in the global economy.

Second, drawing on insights from Weber (2004), I argue that government support for stand-alone solar electrification in Africa also reflects its role as a political safety valve. Energy access is a politically-sensitive issue in many African countries, where inadequate electricity provision often triggers social unrest and political contestation (Ahlborg & Hammar, 2014; De Juan & Wegner, 2019; Kesselring, 2017). OGS electrification mitigates this risk by offering a limited but functional alternative, thereby reducing pressure on governments to expand grid infrastructure.

The case of Ghana's solar electrification efforts illustrates this dynamic. Ghana's National Climate Change Master Plan explicitly promotes renewable energy, including solar, as a means to enhance energy security (Stock et al., 2023). However, while the government publicly justifies OGS expansion as a climate-friendly initiative, it also serves a pragmatic political function: addressing energy poverty without the fiscal and infrastructural burdens of grid expansion. By enabling rural communities to access some level of electricity through decentralized solar technologies, governments can claim progress in electrification without heavy investment in large-scale infrastructure. This reduces social and political pressures, particularly in regions with historically poor access to electricity. In some instances, governments have distributed 'free' OGS products, often as politically-motivated gestures. In Ghana, for instance, the government provided free 500W SHSs to rural communities—not as substitutes for grid electricity but as instruments to manage political expectations for universal access (Boamah & Rothfuß, 2020).

Third, several governments across the continent have actively promoted solar electrification because of its potential to create jobs and stimulate economic activity. By fostering a solar energy sector, governments can reduce employment-related grievances and encourage entrepreneurship. Take the case of Chirwa in Malawi: he uses an SHS to power a television for public viewings in his rural community, generating about MWK12,000 (US$7) each time a football match is broadcast.[32] Similarly, another SHS user, Oswald, explained that 'People pay me MWK50 [US$0.03]' to charge their phones. He fills a gap in a community where mobile phones are common but grid electricity is absent. However, such income-generating activities are structurally unsustainable in the long run due to low barriers to entry, which lead to market saturation (Ojong, 2024). The prospect of profit attracts new entrants, expanding supply and reducing returns. Still, in the short term, rural residents can generate modest income through solar-based micro-enterprises.

Beyond small-scale entrepreneurship, solar energy projects and companies across Africa provide employment both directly and indirectly. Morocco's Noor Ouarzazate Solar Complex, for example, generated about 1000 construction jobs and 60 permanent positions during its operation and maintenance phase (Power Technology, 2020).[33] On a continent with the world's largest youth population and a working-age demographic projected to reach 1.56 billion by 2050 (Bekele-Thomas & Westgaard, 2024), governments already struggling to create jobs cannot ignore the sector's employment potential. Solar electrification, therefore, functions not only as an energy strategy but as a form of labor market policy.

Furthermore, as I show in Chapter 6, the informal solar recycling industry has become a key source of employment across Africa. Solar electrification thus carries indirect economic benefits that contribute to social stability. Nevertheless, as Chapter 6 demonstrates, these gains are often stabilized through women's insertion into household finance and hazardous solar waste work under patriarchal governance.

This form of statecraft also demonstrates not the absence of agency but its bounded nature: African governments appropriate external agendas

[32] The exchange rate used throughout this book is US$1 = MWK1735.

[33] The Noor Ouarzazate Solar Complex is a power plant located 10 km north-east of the city of Ouarzazate, Morocco. As of March 2024, only 831 MW had been installed, far below the 2000 MW that was projected to be installed by 2020 (El Jechtimi & Magid, 2024).

for pragmatic ends, but in ways that frequently reproduce domestic inequalities and global dependence rather than achieving structural transformation. Hence, as noted earlier, solar electrification becomes a revealing arena for the dialectics of structure and agency. Governments can and do maneuver, but their room for maneuver is structured by global financial hierarchies, technological dependence, and the legacies of structural adjustment. The outcome is neither pure imposition nor unfettered sovereignty, but a hybrid configuration in which external pressures and local strategies intersect, opening possibilities for incremental gains while reinscribing the broader architecture of dependency (Moyo & Chikwanha, 2022).

By examining how multilateral development agencies engineer markets, this chapter shows that what appears as technical financial design is, in fact, an architecture of insertion and depletion. Subsequent chapters disclose how these dynamics unfold across households (Chapter 4), landscapes of harm (Chapter 5), and women's everyday negotiations (Chapter 6).

The bounded pragmatism highlighted across these policies and statecraft maneuvers is not confined to ministerial offices or donor negotiations; it descends into everyday life. The same constrained rationalities that guide governments under conditional finance—managing liquidity constraints, deferring risk, and leveraging limited means—reappear within PAYG SHS households that juggle mobile-money payments, informal credit, and community obligations to keep solar lights on. From this angle, state-level bounded pragmatism prefigures the micro-pragmatism of survival explored in the next chapter, where repayment schedules, datafication, and relational debts reproduce insertion and depletion at the scale of domestic economies.

These dynamics also unfold through domestic class politics. As in other contexts (see Boateng & Klopp, 2024; Buur et al., 2012), elite capture and socio-political dynamics shape where solar energy investments land: government-funded SHS programs often concentrate in politically-strategic regions.[34] Therefore, bounded agency is exercised not

[34] For more about elite capture in various contexts across Africa, see, Arnall et al. (2013), Hall and Kepe (2017), Muyengwa et al., (2014), and Wong (2010). For a detailed discussion on

only within global financial hierarchies but through internal social stratifications that privilege local elites and male-dominated enterprises.[35] This dual movement—external dependence intertwined with internal class and gendered divisions—prefigures the argument in Chapter 6, where I show how women's insertion into household finance and the end-of-life work of toxic solar products become essential parts of the quiet infrastructure that stabilizes solar energy accumulation.

But whether these maneuverings amount to structural transformation cannot be judged from statecraft alone. The classed and gendered dynamics discussed earlier are nested within, and often reinforced by, Africa's broader insertion into global trade and production hierarchies. To grasp this larger picture, we must widen the lens to examine whether solar energy's expansion disrupts or reproduces the continent's long-standing dependence on primary commodities.

3.5 Solar Power Boom and the Reproduction of Structural Dependence

As I have emphasized throughout this chapter, dependency (Ghosh, 2019; Velasco, 2002) and world-systems (Klak, 2014; Wallerstein, 2004) literature contend that participation in global markets seldom alters Africa's structural location; rather, it often entrenches it. The expansion of solar energy markets raises a central question: does expanding solar power capacity reposition African economies, or does it replicate unequal exchange under a green guise? I begin by situating solar power economies within this broader structural context.

In the 1970s, soon after many African countries gained independence, Richard Harris (1975: 12) observed that 'as a result of their colonial

social stratification in Africa, including sources of inequality both within African countries and between Africa and the rest of the world, see Obeng-Odoom (2020).

[35] On class, status, and rural–urban dynamics in the SHS sphere in Ghana, see Boamah and Rothfuß (2020: 9).

legacy, the present-day economies of the African countries are characterized by a lop-sided dependence on the export of raw materials and the import of manufactured goods.' More than fifty years after Harris's publication, little has changed. A 2014 report concluded that 'for many African countries, exports are concentrated in a narrow range of primary products that has remained much the same over the past 40 years' (African Center for Economic Transformation, 2014: 27). Between 1995 and 2009, Africa's share of primary commodity exports increased from 72 to 81% (UNDP, 2011: 60), and by 2021, 83% of African countries were still commodity dependent (UNCTAD, 2022).

As shown in Fig. 3.3, Africa remains the world's most commodity-dependent region, accounting for 44.6% of all commodity-dependent countries.[36] Figure 3.4 further shows that in 2023, one product group accounted for at least 50% of merchandise exports in 21 of the 54 African countries.[37] According to the United Nations Conference on Trade and Development's (UNCTAD) classification, these product groups include 'fuels' (petroleum oils and oils obtained from bituminous materials), 'ores and metals,' and 'all food items.' Six countries (Algeria, Angola, Equatorial Guinea, Libya, South Sudan, and Malawi) depended on a single product line for at least 90% of their export earnings. Except for Malawi, which relies on food exports, the others are oil producers. This heavy reliance exposes economies to extreme price volatility, which constrains fiscal capacity and undermines the structural transformation urgently needed across the continent.[38]

This dynamic is emblematic of what dependency theorists termed unequal exchange, where peripheral economies are locked into exporting raw materials while importing higher value manufactured goods (Amin,

---

[36] A country is considered 'commodity-dependent' when a significant portion of its merchandise exports consists primarily of primary commodities such as minerals, ores, metals, fuels, agricultural raw materials, and food (UNCTAD, 2023a).

[37] 54 African countries excluding Sahrawi Arab Democratic Republic, a former Spanish colony annexed by Morocco in 1975, that is recognized by several governments and is a full member of the African Union. That country was excluded due to no data.

[38] Between January 2014 and June 2021, oil prices experienced 37 downturns, averaging a 10% decline, while during the 2007–2009 financial crisis, crude oil prices dropped by over 25% in a single month (Hansen & Gerval, 2022). At the onset of the COVID-19 pandemic, oil prices plummeted by more than 42% (March–April 2020), eroding export revenues, depreciating currencies, and depleting foreign exchange reserves (Hansen & Gerval, 2022).

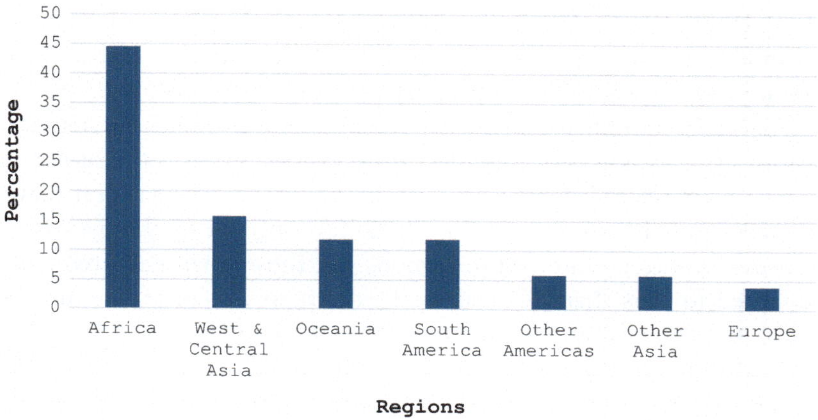

Fig. 3.3 Commodity dependence by region, 2019–2021 (Data sourced from UNCTADStat database. Available at: https://unctadstat.unctad.org/datacentre/ (accessed January 9, 2025))

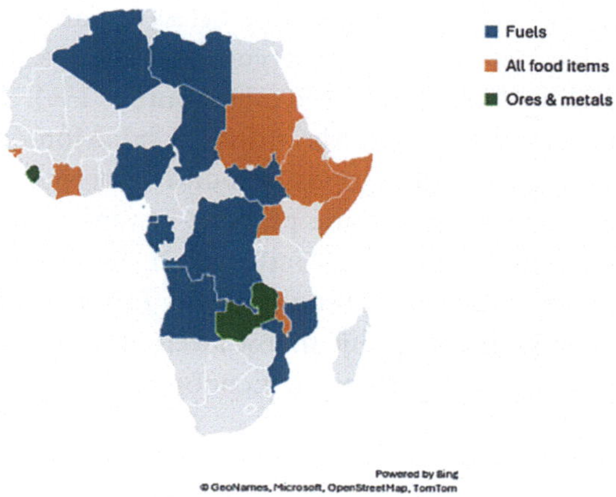

Fig. 3.4 African countries where one product group was at least 50% of merchandise exports in 2023 (Data sourced from UNCTADStat database. Available at: https://unctadstat.unctad.org/datacentre/ (accessed January 9, 2025))

1976, 1990; Frank, 1967). Far from generating endogenous growth, commodity booms reproduce vulnerability by reinforcing Africa's structural location as a supplier of primary goods to the global core.

Whereas commodity dependence reveals one side of Africa's structural position, the weakness of manufacturing exports exposes the other: the absence of industrial depth. An examination of export data on manufactured goods is just as revealing. In 2023, only 44 of the 54 African countries exported manufactured goods. Of those, manufactured goods accounted for less than 20% of total exports in 29 countries. In other words, for 66% of the countries that exported manufactured goods, these products were not a significant part of their export portfolios. This stands in sharp contrast to Asia, the Middle East, Europe, and North America, where manufactured goods dominate: China (94%), South Korea (86%), France (78%), Germany (87%), Israel (81%), Italy (82%), Japan (91%), Singapore (74%), the United States (66%), and the United Kingdom (67%) in the same year.

As shown in Fig. 3.5, there are nonetheless variations across Africa regarding the share of manufactured goods in total exports. In eight countries, including Comoros (51%), Djibouti (51%), Egypt (52%), Eswatini (59%), Lesotho (55%), Mauritius (51%), Morocco (71%), and Tunisia (78%), manufactured goods accounted for more than half of exports.[39] These cases suggest some progress toward manufacturing-led growth, yet they remain exceptions rather than the norm. Even among these relatively industrialized economies, many export products still 'fall short in global competitiveness due to low levels of technology and insufficient innovation' (African Center for Economic Transformation, 2023: 23).

[39] Tunisia's position as a leader in the export of manufactured goods is attributed to the country's economic policies. Specifically, the *Programme de Mise à Niveau* played a crucial role in enhancing firms' competitiveness by revitalizing the industrial sector, which, in turn, facilitated economic diversification through technical support, workforce training, financial incentives, and infrastructure improvements (African Center for Economic Transformation, 2023). Additionally, Tunisia promoted export-oriented policies driven by the private sector, which helped attract trade and investment collaborations, and these efforts integrated the national economy into both 'regional and global manufacturing value chains' (African Center for Economic Transformation, 2023: 19).

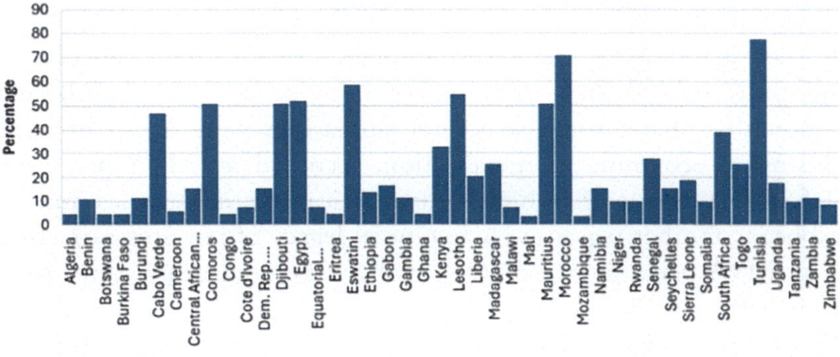

Fig. 3.5 Manufactured goods as share of merchandise exports, 2023 (Data sourced from UNCTADStat database. Available at: https://unctadstat.unctad.org/datacentre/ (accessed January 9, 2025))

In short, the data reveal that despite pockets of progress, Africa's export portfolio remains overwhelmingly oriented toward primary commodities, reinforcing structural dependency and stifling industrial transformation. Low manufacturing value added underscores that even as millions of OGS units circulate, the continent captures little of the higher value industrial returns embedded in those technologies. The very boom that promises energy inclusion simultaneously reproduces a trade structure in which Africa imports finished solar hardware while exporting the raw minerals that feed it, entrenching the unequal exchange at the heart of solar power capitalism.

From a world-systems perspective, the data reveal Africa's continued insertion as a peripheral region within the international division of labor, where the core controls advanced industrial production and captures value (see also Wallerstein, 1974, 2004). Even where manufacturing exports rise modestly, they often exemplify 'dependent development'—industrialization that is externally oriented, technologically shallow, and unable to disrupt structural subordination (Cardoso & Faletto, 2024).

Moreover, a key measure of progress toward a more diversified production structure is the proportion of manufacturing value added (MVA) in a country's gross domestic product (Taylor, 2016). In fact, MVA per capita also serves as a basic benchmark for assessing a country's degree of industrialization (Taylor, 2016). Here again, African countries lag their counterparts in other regions (Table 3.3). Between 2019 and 2021, not only was Africa's MVA the lowest among all regions but it also lagged significantly behind those regions. For instance, in 2021, South Asia, a region that shares some structural similarities with Africa in terms of historical economic marginalization, recorded an MVA per capita of US$494.2, about 58% higher than Africa's. Latin America and the Caribbean, another region with a colonial legacy of resource extraction, had an MVA per capita of $898.2 in the same year, nearly three times Africa's level.

What is even more striking is the contrast between Africa and regions that have achieved a high degree of industrialization. In 2021, East Asia and the Pacific had an MVA per capita of US$3026, nearly ten times Africa's figure, reflecting the successful integration into global manufacturing supply chains of several countries in that region.[40] Even the Middle East, a region that is often viewed through the lens of oil dependency, reported an MVA per capita of US$1888.6, more than six times Africa's.

It is important to note that there are MVA variations across Africa (Fig. 3.6). For example, Mauritius and Eswatini exhibit the highest

Table 3.3 Regional averages of manufacturing value added per capita at constant (2015) US$

	2019	2020	2021
Africa	332.7	306.9	313.6
South Asia	488.5	475.6	494.2
Europe & Central Asia	3603.4	3557.5	3892.5
Middle East	1938.5	1815.1	1888.6
Latin America & Caribbean	946.8	879	898.2
East Asia & Pacific	2790.1	2853	3026
North America	3470.7	3242.2	3381.7

Notes Data sourced from UNIDO's Competitive Industrial Performance Index 2019–2021. Available at: https://stat.unido.org/analytical-tools/cip?country=124 (accessed January 9, 2025)

[40] Of course, there are national variations within the East Asia & Pacific region. For example, in 2021, Australia had an MVA per capita of US$3031, Cambodia (US$219), China (US$3145), Singapore (US$13,783), Mongolia (US$339), Papua New Guinea (US$40), Tonga (US$215), and Vietnam (US$872).

MVA per capita, with figures reaching US$1237 and US$1206 in 2019, respectively.[41] However, by 2021, these values had declined to US$1095 and US$1042, reflecting economic disruptions or structural challenges.[42] Conversely, countries such as Burundi and Gambia report some of the lowest MVA per capita, with figures as low as US$29 and US$13 in 2021, highlighting persistent industrial underdevelopment.

Additionally, there are notable subregional contrasts across the continent. In North Africa, Tunisia and Egypt maintained relatively higher manufacturing value added (MVA) per capita, recording US$515 and US$543 in 2021, respectively. In contrast, Libya exhibited more volatility, with its MVA per capita fluctuating from US$217 in 2019

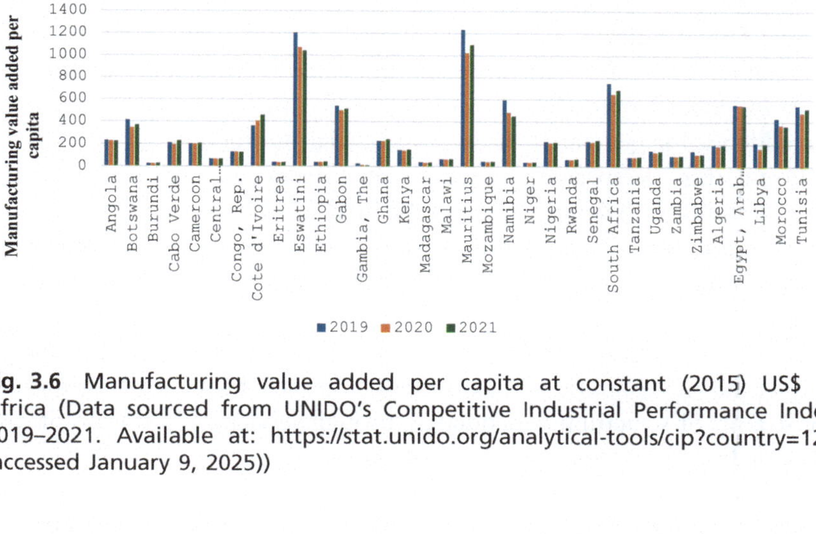

Fig. 3.6 Manufacturing value added per capita at constant (2015) US$ in Africa (Data sourced from UNIDO's Competitive Industrial Performance Index 2019–2021. Available at: https://stat.unido.org/analytical-tools/cip?country=124 (accessed January 9, 2025))

[41] Eswatini's impressive performance is due to its leveraging of the African Growth and Opportunity Act and the Multi-Fiber Act of 2004 to strengthen its domestic textile and garment supply chains, while also expanding its processing capabilities for various non-extractive goods (African Center for Economic Transformation, 2023). Along similar lines, Mauritius transitioned from a sugar cane-dependent, commodity-based economy to a more diversified economic model, with significant growth in processed goods and service industries such as textiles, financial services, and tourism (African Center for Economic Transformation, 2023).

[42] In April 2025, the Trump administration imposed 10% tariff on goods imported from Eswatini (Dlamini, 2025, April 3) and 40% tariff on goods imported from Mauritius (The Voice of Africa, 2025, April 21), which will negatively impact their export of manufactured goods.

to US$168 in 2020 before rebounding to US$206 in 2021.[43] In West Africa, Côte d'Ivoire stood out for its steady increase in MVA per capita (from US$365 in 2019 to US$457 in 2021) suggesting a more robust industrial expansion compared to neighboring economies such as Nigeria, which remained relatively stagnant at around US$216 in 2021. Southern Africa also presents mixed trends. While South Africa, the region's industrial powerhouse, saw a decline from US$753 in 2019 to US$687 in 2021, other countries like Namibia and Botswana experienced sharper contractions, with Namibia's MVA per capita dropping from US$597 to US$449 over the same period.

The persistently low levels of MVA per capita across Africa indicate a structural economic dependency rooted in the continent's continued reliance on the export of primary commodities rather than manufactured goods. This condition, a legacy of colonial economic structures, as noted in Chapter 1, renders African economies highly vulnerable to fluctuations in global commodity prices while preventing them from capturing the added value generated through industrial production.

From this perspective, Africa's dependency is not just a question of trade imbalances; it is embedded in the very architecture of global capitalism, where manufacturing powerhouses capture value while most resource-rich countries remain locked in a cycle of underdevelopment.

Now, some argue that Africa is making important strides in services. While this is accurate, the continent will not break away from the constraints of structural dependency through services. On this point, the African Transformation Report 2014 notes:

> One indicator of progress toward a more diversified production structure is the share of manufacturing value added in GDP [...] Indeed, it appears that Sub-Saharan countries are directly replacing agriculture with services as the largest economic sector without passing through the intermediate phase of industrialization and an expanding manufacturing sector, the experience of almost all successful economies. Moreover, a large

[43] Libya's low and volatile MVA may be attributed to the civil war that has plagued the country for over a decade.

> part of the services sector in many Sub-Saharan countries consists of low-technology and low-value activities. These trends are of great concern, since manufacturing has historically been the main source of technological learning. (African Center for Economic Transformation, 2014: 27)

A shift to low-productivity services, while potentially beneficial in some respects, does not fundamentally disrupt the deep-seated patterns of structural dependence across African economies. Indeed, services alone often fail to offer the kind of robust technology transfers, skill development, and large-scale job creation historically found in manufacturing. By bypassing an intermediate phase of industrialization, African countries may gain a temporary increase in gross domestic product (GDP) contributions from low-productivity service sectors, but they risk missing out on the industrial spillover effects that have catalyzed the long-term economic transformations in other regions. It is now clear that Africa remains trapped in an entrenched condition of structural dependence. So, does the solar power boom on the continent alleviate or deepen this longstanding problem? I respond to this question in the next section.

3.6 Solar Power and the Reproduction of the Resource-Extraction-Export Model

These patterns of dependency are not limited to current trade structures; they also extend into the mineral base that underpins solar technologies themselves. Africa's critical minerals demonstrate how the continent's role in global solar power capitalism is once again framed through extraction rather than transformation. The very materials that make renewable energy possible are drawn disproportionately from African soils, revealing a continuity between fossil and green extractivism.

Africa holds a significant share of the world's critical minerals essential for solar power technologies and the renewable energy sector more broadly, including aluminum, cobalt, and copper. Guinea, for example, possesses nearly one-quarter of the world's bauxite reserves, the primary source of aluminum, which constitutes over 85% of solar photovoltaic components (Mo Ibrahim Foundation, 2022). Southern African countries, including South Africa, accounted for 80% of global chromium ore exports in 2020, a mineral vital for concentrated solar power, nuclear, geothermal, wind, and hydropower technologies (ibid.). The Democratic Republic of Congo (DRC) supplies more than 70% of the world's cobalt, a crucial mineral for lithium-ion batteries used in electric vehicles and energy storage (IMF, 2024; Mo Ibrahim Foundation, 2022). In 2021, Zambia and the DRC together produced 12% of the world's copper, a metal used across renewable energy systems from solar to wind and hydro (n, 2022). Between 2016 and 2020, Zambia was the world's largest exporter of unrefined copper (ibid.). Madagascar, Mozambique, and Tanzania collectively account for one-fifth of global graphite reserves, another critical input for lithium-ion batteries (Mo Ibrahim Foundation, 2022). Finally, South Africa, Côte d'Ivoire, Ghana, and Gabon produce roughly 60% of the world's manganese, essential for concentrated solar power (ibid.).

These minerals, which are vital to Africa's ongoing solar power boom, are not contributing to the structural transformation of the countries that produce them. The IMF's 2024 *Regional Economic Outlook* was blunt:

> Most sub-Saharan African countries export critical minerals primarily in their raw form, which tend to be lower value added than processing activities. The economic disparity is evident in a simple market value comparison: raw bauxite fetches a modest $65 per ton, whereas its processed counterpart, aluminum, commands a hefty $2,335 per ton as of end 2023 [...] The Democratic Republic of the Congo, accounting for 74 percent of global cobalt mining, sends 97 percent of its cobalt exports, mostly unprocessed, to China. Similarly, over 1,000 trucks are estimated to daily transport low-concentration, rock-form lithium from Zimbabwean mines to distant African ports, destined for China instead of undergoing local processing (Goldman Sachs, 2023). This approach starkly contrasts with China and Indonesia, who have prioritized mineral processing for higher economic returns. (IMF, 2024: 3–4)

It is evident that although the rise of solar power has increased the demand for minerals essential to the industry, African countries that produce these critical resources have not leveraged this opportunity to achieve structural economic transformation. In other words, the solar energy boom is simply reinforcing resource extraction, perpetuating a pattern that has persisted for decades.

The export structures of Zambia and the DRC reveal a striking continuation of the traditional resource-extraction-export model, reinforcing the economic dependency of these countries on unprocessed mineral exports (Fig. 3.7). According to data from the UNCTADStat database, ores and metals accounted for 67% of Zambia's total exports and 80% of the DRC's exports in 2023. Meanwhile, manufactured goods, i.e., products that typically yield greater economic value and contribute to industrialization, constituted only 12% of Zambia's exports and 16% of the DRC's. This imbalance underscores the failure to develop domestic processing industries that could capture more value from these resource endowments..

As dependency theory would predict, the solar energy surge has created a new 'green extractivism' that reproduces old hierarchies under the guise of sustainability.[44] African states provide the raw minerals essential to solar power capitalism, while the value-added processing and technological upgrading occur elsewhere, ensuring that Africa remains structurally dependent on external actors for industrial transformation.

3.6.1 A Market for Solar Power, but Not a Leading Manufacturer of Solar Goods

Earlier, I mentioned that millions of solar products have been sold in Africa. However, despite the continent's vast solar energy potential, it remains overwhelmingly dependent on foreign manufacturers for its solar energy needs. Recent trade data on solar photovoltaic (PV) imports reveal a stark reality: Africa imports almost all its PV components, including photovoltaic cells (both assembled and unassembled), direct current (DC) and alternating current (AC) generators, and complete solar panel systems. The trade figures are telling.

---

[44] For a detailed discussion about green extractivism, see Bruna (2023). On the coloniality of green extractivism, see Andreucci et al. (2023).

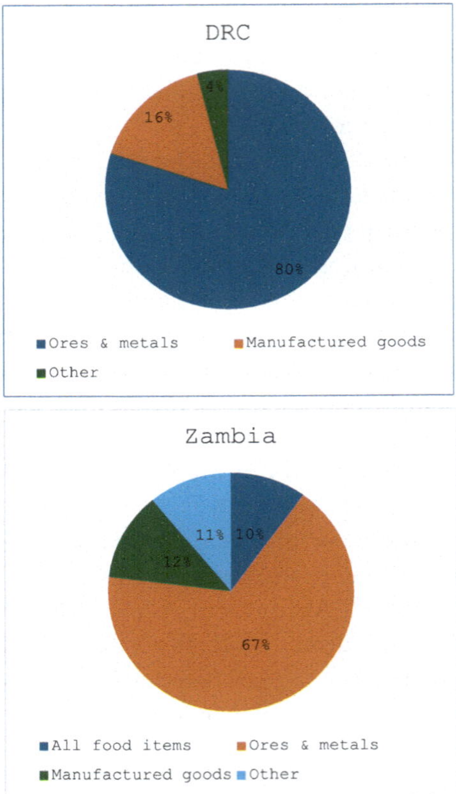

Fig. 3.7 Export structure of Zambia and the Democratic Republic of Congo, 2023 (Data sourced from UNCTADStat database. Available at: https://unctadstat.unctad.org/datacentre/ (accessed January 9, 2025))

In 2023, African countries imported millions of PV cells and modules, with South Africa alone receiving nearly 48.7 million units of photovoltaic cells assembled in modules or panels. Other African countries such as Mauritius, Morocco, and Egypt also recorded substantial imports, with quantities in the tens of thousands, further highlighting the continent's heavy reliance on external suppliers. Unassembled PV cells followed a similar pattern: South Africa imported over 374,000 units in 2023, while Mauritius, Côte d'Ivoire, and Namibia also recorded smaller imports (Table 3.4). This trade pattern is significant because it shows that even basic assembly processes remain largely offshore. Africa is also a key consumer of solar generators. In 2023 alone, South Africa, Mauritius, and Nigeria imported major quantities of both DC and AC photovoltaic generators. Mauritius imported over 886,000 units of photovoltaic AC generators, while South Africa imported 2139, illustrating the depth of reliance on imported systems (Table 3.4). Other countries such as Uganda, Senegal, and Cameroon followed similar trends, albeit with smaller volumes.

In recent years, Africa has witnessed the emergence of local solar PV manufacturers seeking to reduce this dependency and meet domestic energy demand. Examples include Auxano Solar Nigeria Limited, ARTsolar in South Africa, and Faso Energy in Burkina Faso. Auxano Solar Nigeria Limited, established in 2016 as Nigeria's first privately owned solar panel assembly facility, 'envisions supplying at least 20% of the solar modules used in Nigeria' (Sustainable Energy for All, 2024: 2), a reminder that imported panels still dominate the market.[45] South Africa hosts ARTsolar, a 100% locally-owned PV module manufacturer, while Faso Energy in Burkina Faso, founded in 2020, produces between 60 and 100 photovoltaic panels per day.[46] These efforts signal emerging industrial agency, but their limited scale and capacity reveal how the continent's position within global solar power capitalism remains largely that of a consumer rather than a producer.

Despite these limited efforts, the broader African solar energy market continues to rely heavily on imported solar products. The presence of a few local manufacturers, while promising, remains modest in scale compared to the vast import volumes. Globally, the top solar panel manufacturers are located in North America and Asia. These include Canadian Solar (Canada), SunPower (United States), First Solar (United States), Trina Solar (China), Jinko Solar (China), and Hanwha Q CELLS (South Korea) (UNCTAD, 2023b). Likewise, leading firms in the concentrated solar power industry—BrightSource Energy (United

[45] It also deals with the repair and maintenance of inverters. More information about Auxano Solar Nigeria Ltd. can be found on their website, https://auxanosolar.com/.

[46] For detailed information about ARTsolar, visit their website, https://artsolar.net/.

Table 3.4 Leading importers of photovoltaic products in Africa

Photovoltaic cells assembled in modules or made up into panels			Photovoltaic cells not assembled in modules or made up into panels			Photovoltaic AC generators			Photovoltaic DC generators, of an output >50 W		
Country	Year	Quantity	Country	Year	Quantity	Country	Year	Quantity	Country	Year	Quantity
South Africa	2023	48,705,414 units	South Africa	2023	374,757 units	Mauritius	2023	886,252 units	Nigeria	2023	5800 tons
Mauritius	2023	37,498 units	Mauritius	2023	3302 units	South Africa	2023	2139 units	South Africa	2023	751 units
Morocco	2023	30,568 tons	Côte d'Ivoire	2023	1635 tons	Nigeria	2023	624 tons	Mauritius	2023	264 units
Egypt	2023	27,300 tons	Namibia	2023	1209 tons	Uganda	2023	496 tons	Somalia	2023	215 tons
Namibia	2023	17,215 tons	Senegal	2023	528 tons	Senegal	2023	443 tons	Chad	2023	132 tons
Burkina Faso	2023	12,762 tons	Burkina Faso	2023	396 tons	Madagascar	2023	361 tons	Cameroon	2023	105 tons
Kenya	2023	10,264 tons	Uganda	2023	394 tons	Somalia	2023	244 tons	Mali	2023	103 tons
Madagascar	2023	8935	Nigeria	2023	393 tons	Cameroon	2023	130 tons	Guinea	2023	96 tons
Uganda	2023	6517 tons	Malawi	2023	256 tons	Chad	2023	129 tons	Botswana	2023	45 tons
Tunisia	2023	4477 tons	Botswana	2023	206 tons	Tunisia	2023	81 tons	Ethiopia	2023	43 tons

Notes Data sourced from International Trade Centre (ITC) Trade Map database. Available at: https://www.trademap.org/Index.aspx (accessed January 8, 2025)

States), NextEra Energy Resources (United States), Ibereolica Group (Spain), Abengoa Solar, S.A. (Spain), and ENGIE (France)—are all headquartered in the global North (UNCTAD, 2023b). Africa, by contrast, has established itself as a mass consumer of solar energy technologies but not as a significant producer.

The absence of large-scale value-added manufacturing capacity means that African economies capture little value from the solar power industry beyond end-user consumption. The continent's solar energy expansion, therefore, reproduces what Shivji (2009: 59) calls structural disarticulation: 'the disarticulation between the structure of production and the structure of consumption: what is produced is not consumed and what is consumed is not produced.' From this angle, solar energy growth, while symbolically tied to sustainability and innovation, materially deepens the same structure of dependent integration that has long defined Africa's position in the world economy.

3.7 Concluding Remarks

This chapter has shown that solar power capitalism is not a spontaneous market evolution but a state-mediated project that converts surplus capital into new asset classes while repositioning African economies as testing grounds for concessional finance. Through what I have called *bounded pragmatism*, African governments negotiate limited agency within structures of dependency they did not design but must inhabit. These dynamics lay bare how insertion occurs not only at the household or community level but also at the scale of the state, which becomes an intermediary of global accumulation. With this in mind, I now follow these macro-logics downward, showing how the same circuits of finance and obligation materialize within households through everyday debt, repayment, and social discipline.

References

African Center for Economic Transformation. (2014). *African transformation report 2014: Growth with depth*. African Center for Economic Transformation.

African Center for Economic Transformation. (2023). *African transformation index*. African Center for Economic Transformation.

Ahlborg, H., & Hammar, L. (2014). Drivers and barriers to rural electrification in Tanzania and Mozambique: Grid-extension, off-grid, and renewable energy technologies. *Renewable Energy, 61*, 117–124.

Akon Lighting Africa. (2016). *Akon Lighting Africa: Providing electricity to African villages by solar energy*.

Ambrose, J. (2019, July 25). UK solar power pioneer Solarcentury profit grows 860% in a year. *The Guardian*. Retrieved on February 23, 2025 from https://www.theguardian.com/business/2019/jul/25/uk-solar-power-pioneer-solarcentury-profit-grows-860-in-a-year

Amin, S. (1976). *Unequal development: An essay on the social formations of peripheral capitalism*. Monthly Review Press.

Amin, S. (1990). *Delinking: Towards a polycentric world*. Zed Books.

Andreucci, D., López, G. G., Radhuber, I. M., Conde, M., Voskoboynik, D. M., Farrugia, J. D., & Zografos, C. (2023). The coloniality of green extractivism: Unearthing decarbonisation by dispossession through the case of nickel. *Political Geography, 107*, 102997.

Arnall, A., Thomas, D. S., Twyman, C., & Liverman, D. (2013). NGOs, elite capture and community-driven development: Perspectives in rural Mozambique. *The Journal of Modern African Studies, 51*(2), 305–330.

Arrighi, G. (1990). The developmentalist illusion: A reconceptualization of the semiperiphery. In W. Martin (Ed.), *Semiperipheral states in the world-economy* (pp. 11–42). Greenwood Press.

Azuri Group. (2019, June 4). *Solar pioneer Azuri Technologies announces $26 million equity investment*. Retrieved on February 24, 2025 from https://www.azuri-group.com/industry-insights/26-million-investment

Baptista, I., & Plananska, J. (2017). The landscape of energy initiatives in sub-Saharan Africa: Going for systemic change or reinforcing the status quo? *Energy Policy, 110*, 1–8.

Bekele-Thomas, N., & Westgaard, S. (2024, October 4). Unlocking the potential of Africa's youth. *Africa Renewal*. Retrieved on February 22, 2025 from https://www.un.org/africarenewal/magazine/october-2024/unlocking-potential-africa%E2%80%99s-youth

Bernard, T. (2012). Impact analysis of rural electrification projects in sub-Saharan Africa. *The World Bank Research Observer, 27*(1), 33–51.

Berndt, C., & Boeckler, M. (2009). Geographies of circulation and exchange: Constructions of markets. *Progress in Human Geography, 33*(4), 535–551.

Beyer, M. (2024, April 3). *Cameroon exempts imported PV components from customs duties*. Retrieved on February 18, 2025 from https://www.pv-magazine.com/2024/04/03/cameroon-exempts-imported-pv-components-from-customs-duties/

Bhamidipati, P. L., & Hansen, U. E. (2021). Unpacking local agency in China-Africa relations: Frictional encounters and development outcomes of solar power in Kenya. *Geoforum, 119*, 206–217.

Bhattacharyya, S. C. (2012). Electrification experiences from sub-Saharan Africa. *Rural electrification through decentralised off-grid systems in developing countries* (pp. 131–156). Springer.

BII [British International Investment]. (2023). *BII & Stanbic Bank Kenya in double commitment to off-grid solar energy company Sun King*. Retrieved on February 9, 2025 from https://www.bii.co.uk/en/news-insight/news/bii-stanbic-bank-kenya-in-double-commitment-to-off-grid-solar-energy-company-sun-king/#:~:text=Growing%20with%20early%2Dstage%20funding,General%20Atlantic%2C%20M%26G%20and%20Leapfrog

Bishop, M., & Green, M. (2015). Philanthrocapitalism Rising. *Society, 52*, 541–548.

Boamah, F., & Rothfuß, E. (2020). 'Practical recognition' as a suitable pathway for researching just energy futures: Seeing like a 'modern' electricity user in Ghana. *Energy Research & Social Science, 60*, 101324.

Boateng, F. G., & Klopp, J. M. (2024, December). Urbanization, legacies of elite capture, and multi-dimensional exclusions in Ghana: Towards just housing and neighborhood policies in African cities. In *Urban forum* (Vol. 35, No. 4, pp. 499–528). Springer.

Bradbrook, A. J. (1992). Energy conservation legislation for industry. *Journal of Energy & Natural Resources Law, 10*(1–4), 145–163.

Brew-Hammond, A. (1996). The electricity supply industry in Ghana: Issues and priorities. *Africa Development/Afrique et Développement, 21*(1), 81–98.

Budabin, A. C., Rasmussen, L. M., & Richey, L. A. (2017). Celebrity-led development organisations: the legitimating function of elite engagement. *Third World Quarterly, 38*(9), 1952–1972.

Bruna, N. (2023). *The rise of green extractivism: Extractivism, rural livelihoods and accumulation in a climate-smart world*. Routledge.

Burger, A. (2019, March 26). The solar and renewables-fuelled utility of the future may be evolving in Sub-Saharan Africa. *Solar Magazine.* Retrieved on February 18, 2025 from https://solarmagazine.com/the-solar-and-renewables-fueled-utility-of-the-future-may-be-evolving-in-sub-saharan-africa/

Buur, L., Tembe, C. M., & Baloi, O. (2012). *Mozambique synthesis analysis: Between pockets of efficiency and elite capture*. Institut for Internationale Studier/Dansk Center for Internationale Studier og Menneskerettigheder.

Cardoso, F. H., & Faletto, E. (2024). *Dependency and development in Latin America*. University of California Press.

Carroll, T., & Jarvis, D. S. (2014). Introduction: Financialisation and development in Asia under late capitalism. *Asian Studies Review, 38*(4), 533–543.

Ćetković, S., & Buzogány, A. (2016). Varieties of capitalism and clean energy transitions in the European Union: When renewable energy hits different economic logics. *Climate Policy, 16*(5), 642–657.

Charo, R. (1986). Alternative energy power production: The impact of the Public Utility Regulatory Policy Act. *Columbia Journal of Environmental Law, 11*, 447.

Cherrington, R., Goodship, V., Longfield, A., & Kirwan, K. (2013). The feed-in tariff in the UK: A case study focus on domestic photovoltaic systems. *Renewable Energy, 50*, 421–426.

Clifford, M. L. (2015). *The greening of Asia: The business case for solving Asia's environmental emergency.* Columbia University Press.

Cooper, F. (2002). *Africa since 1940: The past of the present*. Cambridge University Press.

Cross, J. (2019). The solar good: Energy ethics in poor markets. *Journal of the Royal Anthropological Institute, 25*(S1), 47–66.

Cross, J., & Neumark, T. (2021). Solar power and its discontents: Critiquing off-grid infrastructures of inclusion in East Africa. *Development and Change, 52*(4), 902–926.

Cudahy, R. D. (1995). PURPA: The intersection of competition and regulatory policy. *Energy Law Journal, 16*, 419.

Danaher, K. (Ed.). (1994). *50 years is enough: The case against the World Bank and the International Monetary Fund*. South End Press.

Davies, T. (2018). Toxic space and time: Slow violence, necropolitics, and petrochemical pollution. *Annals of the American Association of Geographers, 108*(6), 1537–1553.

Daystar Power. (2021, July 6). *Daystar Power secures $20 million facility from IFC*. Retrieved on February 18, 2025 from https://www.daystar-power.com/news-post/daystar-power-secures-20-million-facility-from-ifc

De Juan, A., & Wegner, E. (2019). Social inequality, state-centered grievances, and protest: Evidence from South Africa. *Journal of Conflict Resolution, 63*(1), 31–58.

Diouf, B., & Miezan, E. (2021). The limits of the concession-led model in rural electrification policy: The case study of Senegal. *Renewable Energy, 177*, 626–635.

Dlamini, Z. M. (2025, April 3). Eswatini says 10% tariffs imposed by the United States will negatively impact exporters, expresses commitment to maintain international trade relations. *Swaziland News.* Retrieved on July 7, 2025 from https://www.swazilandnews.co.za/fundza.php?nguyiphi=8747

Dos Santos, T. (2019). The structure of dependence. In *The gap between rich and poor* (pp. 95–104). Routledge.

EDCL [Energy Development Corporation Limited] (2023, July). A concept note on the Rwanda National Electrification Plan (NEP): 2023 revision. Government of Rwanda. https://www.reg.rw/fileadmin/user_upload/Concept_Note_NEP_Revision__July_2023.pdf

Eichler, R. (2006). *Can "pay for performance" increase utilization by the poor and improve the quality of health services?* Discussion paper for the first meeting of the Working Group on Performance-Based Incentives (p. 5). Center for Global Development.

Ekström, S. B., & Eriksson, F. S. (2018). *Solar home systems in rural Zambia.* Retrieved on February 02, 2025 from https://www.diva-portal.org/smash/get/diva2:1246336/FULLTEXT01.pdf

El Jechtimi, A., & Magid, P. (2024, March 24). *Breakdown at Saudi solar plant in Morocco costs firm $47 million.* Reuters. Retrieved on February 06, 2025 from https://www.reuters.com/business/energy/breakdown-saudi-solar-plant-morocco-costs-firm-47-million-2024-03-24/

Energy Sector Management Assistance Program (ESMAP). (2005). *Decentralized rural electrification project in Cameroon.* World Bank. Retrieved on February 02, 2025 from https://openknowledge.worldbank.org/server/api/core/bitstreams/884c5b7e-bbbb-53c0-bea0-d14aa8a35c27/content

Energy Sector Management Assistance Program (ESMAP). (2024). *Off-grid solar market trends report 2024*. World Bank. Retrieved on February 18, 2025 from https://www.esmap.org/sites/default/files/esmap-files/2024-Off-Grid-Solar-Market-Trends-Report.pdf

Esper, H., London, T., & Kanchwala, Y. (2013). *Access to clean lighting and its impact on children: An exploration of SolarAid's SunnyMoney*. University of Michigan.

European Investment Bank [EIB]. (2018). *EU bank announces new financing to strengthen access to energy in Africa via the development of off-grid solar systems*. Retrieved on February 18, 2025 from https://www.eib.org/en/press/all/2018-077-eu-bank-announces-new-financing-to-strengthen-access-to-energy-in-africa-via-the-development-of-off-grid-solar-systems

Evans, P. B. (1979). *Dependent development: The alliance of multinational, state, and local capital in Brazil*. Princeton University Press.

Fehrenbacher, K. (2014). How M-KOPA unlocked pay-as-you-go solar in rural Kenya. GIGACOM. Retrieved November 2, 2021, from https://gigaom.com/2014/04/10/how-m-kopa-unlocked-pay-as-you-go-solar-in-rural-kenya/

Financial Post. (2019, August 28). Mitsubishi leads funding for off-grid power firm BBOXX's expansion. *Financial Post.* Retrieved on February 02, 2025 from https://financialpost.com/pmn/business-pmn/mitsubishi-leads-funding-for-off-grid-power-firm-bboxxsexpansion

First Solar. (2009). *Annual report 2008*. First Solar.

Foucault, M. (1991). *The Foucault effect: Studies in governmentality*. University of Chicago Press.

Frank, A. G. (1967). *Capitalism and underdevelopment in Latin America* (Vol. 93). NYU Press.

Friedman, T. (2008). *Hot, flat, and crowded: Why we need a green revolution—And how it can renew America*. Farrar, Straus, and Giroux.

Frondel, M., Ritter, N., et al. (2008). Germany's solar cell promotion: Dark clouds on the horizon. *Energy Policy, 36*, 4198–4204.

Ghosh, B. N. (2019). *Dependency theory revisited*. Routledge.

Global Off-Grid Lighting Association (GOGLA). (2018). *Global off-grid solar market report: Semi-annual sales and impact data*. Global Off-Grid Solar Energy Industry Association. Retrieved on February 06, 2025 from https://www.gogla.org/global-off-grid-solar-market-report

GOGLA. (2024). *Semi-annual global off-grid solar market reports: Insights from the January–June 2024 sales and impact data* [Public report]. Retrieved on February 06, 2025 from https://gogla.org/reports/semi-annual-solar-market-report/insights-from-the-january-june-2024-sales-and-impact-data/

Goldman Sachs. (2023, April 27). *Global metals & mining: Direct lithium extraction—A potential game changing technology*. Goldman Sachs Research.

Goodnight, J. A., & King, S. T. (1978). *Role of government in solar energy development: A view from the Northwest* (No. PNL-2784). Battelle Pacific Northwest Labs.

Gouchoe, S., Everette, V., & Haynes, R. (2002). *Case studies on the effectiveness of state financial incentives for renewable energy* (No. NREL/SR-620-32819). National Renewable Energy Laboratory (NREL).

Hall, R., & Kepe, T. (2017). Elite capture and state neglect: New evidence on South Africa's land reform. *Review of African Political Economy, 44*(151), 122–130.

Hanafusa, R., & Sese, S. (2019). Home solar business brightens in Africa. *Nikkei Asia.* Retrieved on February 06, 2025 from https://asia.nikkei.com/Business/Business-trends/Home-solar-business-brightens-in-Africa

Hansen, J., & Gerval, A. (2022, September 26). *Sub-Saharan Africa's reliance on oil exports leads to decline in agricultural imports during pandemic.* Amber Waves. U.S. Department of Agriculture, Economic Research Service. Retrieved on February 22, 2025 from https://www.ers.usda.gov/amberwaves/2022/september/sub-saharan-africa-s-reliance-on-oil-exports-leads-to-decline-in-agricultural-imports-during-pandemic

Harper, P. (2021). Spotlight on off-grid solar energy. *Africa Outlook Magazine, 91*(8–9), 14–15.

Harris, R. (Ed.). (1975). *The political economy of Africa.* Schenkman Publishing Company.

Hart, S. L. (1983). The Federal Photovoltaics Utilization Program: An evaluation and learning framework. *Policy Sciences, 15*, 325–343.

Harvey, D. (2014). *Seventeen contradictions and the end of capitalism.* Oxford University Press.

Hoag, H. (2003). *Designing the Delta: A history of water and development in the Lower Rufiji River Basin, Tanzania, 1945–1985.* UMI Dissertation Services.

International Monetary Fund. (2024). Digging for opportunity: Harnessing Sub-Saharan Africa's wealth in critical minerals. In *Regional economic outlook: Sub-Saharan Africa, April 2024* (Analytical Note). International Monetary Fund. Retrieved on February 19, 2025 from https://www.imf.org/-/media/Files/Publications/REO/AFR/2024/April/English/MineralsNote.ashx

Issawi, C. (1978). The 1973 oil crisis and after. *Journal of Post Keynesian Economics, 1*(2), 3–26.

Jacobsson, S., Sandén, B., & Bångens, L. (2004). Transforming the energy system: The evolution of the German technological system for solar cells. *Technology Analysis & Strategic Management, 16*(1), 3–30.

Jones, G., & Bouamane, L. (2012). Power from sunshine: A business history of solar energy (Harvard Business School Working Paper No. 12-105).

Keane, J. (2014). *Pico-solar electric systems: The Earthscan expert guide to the technology and emerging market.* Earthscan.

Kesselring, R. (2017). The electricity crisis in Zambia: Blackouts and social stratification in new mining towns. *Energy Research & Social Science, 30*, 94–102.

Kinally, C., Antonanzas-Torres, F., Podd, F., & Gallego-Schmid, A. (2023). Solar home systems in Malawi: Commercialisation, use and informal waste management. *Sustainable Production and Consumption, 42*, 367–379.

Kiryakova, E., Borodyna, O., Nadin, R., Howe, L., & Cao, Y. (2025). *China's evolving role in Africa's energy transition: Overseas trade and investment in Kenya, Mozambique and South Africa* (ODI Global Report). ODI Global. Retrieved on August 17, 2025 from https://media.odi.org/documents/240331_China_Africa_Energy_Transition_v4.pdf

Klak, T. (2014). World-systems theory: Core, semi-peripheral, and peripheral regions. In *The companion to development studies* (pp. 143–148). Routledge.

Kranefeld, R. (2019). *Beyond the grid: Post-network energy provision in Rwanda.* Forum Humangeographie.

Langevin, M., Brunet-Bélanger, A., & Lefèvre, S. A. (2023). Financialization through payment infrastructure: The philanthrocapitalism of the Mastercard foundation. In *Financializations of development* (pp. 141–154). Routledge.

Lauber, V., & Metz, L. (2004). Three decades of renewable energy policies in Germany. *Energy and Environment, 15*, 599–623.

Leggett, J. (2019). *SolarAID: An updated history in pictures and charts to mark the sale of our 2 millionth solar light.* SolarAid.

Lema, R., Bhamidipati, P. L., Gregersen, C., Hansen, U. E., & Kirchherr, J. (2021). China's investments in renewable energy in Africa: Creating co-benefits or just cashing-in? *World Development, 141*, 105365.

Lighting Global. (2016). *Off-grid solar market trends report 2016.* Bloomberg New Energy Finance; Lighting Global; World Bank; GOGLA.

Lighting Global. (2018). *Off-grid solar market trends report 2018: January 2018.* Dalberg Advisors; International Finance Corporation.

Lighting Global. (2020). *Off-grid solar: Market trends report 2020.* World Bank; GOGLA; ESMAP.

Madrigal, A. (2011). *Powering the dream: The history and promise of green technology.* Da Capo.

Martinot, E., Cabraal, A., & Mathur, S. (2001). World Bank/GEF solar home system projects: Experiences and lessons learned 1993–2000. *Renewable and Sustainable Energy Reviews, 5*(1), 39–57.

Mawhood, R. K., & Gross, R. (2014). Institutional barriers to a "perfect" policy: A case study of the Senegalese Rural Electrification Plan. *Energy Policy, 73*, 480–490.

Mbembe, A. (2020). *Necropolitics*. Duke University Press.

McGoey, L. (2012). Philanthrocapitalism and its critics. *Poetics, 40*(2), 185–199.

Mchome, E. L. (2022). *Blackout blues: A socio-cultural history of vulnerable electricity networks and resilient users in Dar es Salaam, 1920–2020*. Technische Universität Darmstadt.

Menga, F., Rusca, M., & Alba, R. (2023). Philantrocapitalism and the re-making of global water charity. *Geoforum, 144*, 103788.

MEMD [Ministry of Energy and Mineral Development]. (2023). The energy policy for Uganda 2023. Government of Uganda. https://nrep.ug/document/energy-policy-for-uganda-2023/

Multilateral Investment Guarantee Agency. (2021). *Investment guarantee guide: Insuring investments ensuring opportunities.* Retrieved on February 07, 2025 from https://www.miga.org/sites/default/files/2022-04/MIGA%20Investment%20Guarantee%20Guide%202021%20Digital%20Version%20v3.pdf

Ministry of Infrastructure. (2016). *Rural electrification strategy.* Republic of Rwanda.

Mo Ibrahim Foundation. (2022). *Africa's critical minerals: Africa at the heart of a low-carbon future.* Retrieved on February 26, 2025 from https://mo.ibrahim.foundation/sites/default/files/2022-11/minerals-resource-governance.pdf

Moyo, T., & Chikwanha, A. B. (2022). African agency for development: Progress, contradictions, and complexities in the era of globalisation. *International Journal of African Renaissance Studies, 17*(1), 21–40.

Munro, P., Jacome, V., & Samarakoon, S. (2022). Off-grid enterprise: A critical history of small-scale off-grid solar in Sub-Saharan Africa. In N. Ojong (Ed.), *Off-grid solar electrification in Africa: A critical perspective* (pp. 25–64). Springer.

Muyengwa, S., Child, B., & Lubilo, R. (2014). Elite capture: A comparative case study of meso-level governance in four southern Africa countries. In *Adaptive cross-scalar governance of natural resources* (pp. 179–202). Routledge.

Newton, J. (1989). *Uncommon friends: Life with Thomas Edison, Henry Ford, Harvey Firestone, Alexis Carrel, and Charles Lindbergh.* Mariner.

Nyerere, J. K. (1962). *Ujamaa: The basis of African socialism*. Tanganyika Standard Limited.
Nyerere, J. K. (1971). African socialism: Ujamaa in practice. *The Black Scholar, 2*(6), 2–7.
Nyerere, J. K. (1987). Ujamaa: The basis of African socialism. *The Journal of Pan African Studies, 1*(1), 4–11.
Obeng-Odoom, F. (2020). *Property, institutions, and social stratification in Africa*. Cambridge University Press.
Obeng-Odoom, F. (2024). Oil cities, industrial policy and sustainable development: A new historical approach. *Environment & Urbanization, 36*(2), 478–486.
Ockwell, D., & Byrne, R. (2017). *Sustainable energy for all: Innovation, technology, and pro-poor green transformations*. Routledge.
Öhman, M. B. (2007). *Taming exotic beauties: Swedish hydro power constructions in Tanzania in the era of development assistance, 1960s–1990s* (Doctoral dissertation, KTH).
Ojong, N. (2018). Trust, cultural norms and financial institutions in rural communities: The case of Cameroon. *Review of Social Economy, 76*(1), 19–42.
Ojong, N. (2022). Off-grid solar electrification on the rise in Africa, but where to? In N. Ojong (Ed.), *Off-grid solar electrification in Africa: A critical perspective* (pp. 349–368). Springer.
Ojong, N. (2024). Solar power and the struggle for electrification in Africa. *Current History, 123*(853), 175–180.
Ojong, N., & Simba, A. (2020). Trust-building mechanisms in group-based microfinance: A Cameroonian perspective. *Forum for Social Economics, 49*(2), 180–201.
Oxman, A. D., & Fretheim, A. (2009). Can paying for results help to achieve the Millennium Development Goals? Overview of the effectiveness of results-based financing. *Journal of Evidence-Based Medicine, 2*(2), 70–83.
Pacific Solar. (2001). *Annual review 2000*. Pacific Solar.
Pedersen, R. H., & Poncian, J. (2024). The political economy of energy transitions in Africa: Coalitions, politics and power in Tanzania. *Energy Research & Social Science, 117*, 103712.
Perlin, J. (2002). *From space to earth: The story of solar electricity*. Harvard University Press.
Polanyi, K. (1944). *The great transformation*. Beacon Press.

Power, M., & Kirshner, J. (2019). Powering the state: The political geographies of electrification in Mozambique. *Environment and Planning C: Politics and Space, 37*(3), 498–518.

Power Technology. (2020, March 6). *Noor Ouarzazate solar complex*. Retrieved on February 09, 2025 from https://www.power-technology.com/projects/noor-ouarzazate-solar-complex/

Price, D. (2016). Banking on the poor: Using the off-grid solar revolution to unlock credit for low-income customers in Africa. In D. Bank (Ed.), *Making markets work for the poor* (pp. 13–15). Bill Gates Foundation.

Quinn, J. (1985, April 2). Maverick using new technology: Arco Solar official left to build own photovoltaic cells. *Los Angeles Times.* Retrieved on February 12, 2025 from https://www.latimes.com/archives/la-xpm-1985-04-02-fi-19489-story.html

Rai, S. M., Hoskyns, C., & Thomas, D. (2014). Depletion: The cost of social reproduction. *International Feminist Journal of Politics, 16*(1), 86–105.

Ricaud, A. (2013). Électricité photovoltaïque: Politiques publiques et conséquences économiques: Les choix français dans le contexte international (1973–2013). *Annales historiques de l'électricité, 1*, 111–131.

Roessner, J. D. (1982). Government-industry relationships in technology commercialization: The case of photovoltaics. *Solar Cells, 5*(2), 101–134.

RoL [Republic of Liberia]. (2024). Executive Order No. 134: Renewal of executive order on off-grid solar renewable energy products [Executive order]. The Executive Mansion. https://www.emansion.gov.lr/sites/default/files/documents/Executive%20Order%20No.%20134%20-%20Renewal%20of%20Executive%20Order%20on%20off-grid%20solar%20RE%20products_0.pdf

Rowlands, I. H. (2005). Envisaging feed-in tariffs for solar photovoltaic electricity: European lessons for Canada. *Renewable and Sustainable Energy Reviews, 9*(1), 51–68.

Sadasivam, B. (1997). The impact of structural adjustment on women: A governance and human rights agenda. *Human Rights Quarterly, 19*(3), 630–665.

Samarakoon, S., Bartlett, A., & Munro, P. (2021). Somewhat original: Energy ethics in Malawi's off-grid solar market. *Environmental Sociology, 7*(3), 164–175.

Scavo, J. M. (2015). *False dawn of a solar age: A history of solar heating and power during the energy crisis, 1973–1986* (Doctoral dissertation). University of California, Davis.

Scott, R. (1978). An economic comparison of three technologies: Photovoltaics, nuclear power, co-generating engines. *Environment: Science and Policy for Sustainable Development, 20*(3), 11–12.

Sguazzin, A. (2025). *World Bank's power access drive wins backing from 27 African nations.* Retrieved on February 07, 2025 from https://financialpost.com/pmn/business-pmn/world-banks-power-access-drive-wins-backing-from-27-african-nations

Shah, S. S., & Asghar, Z. (2023). Dynamics of social influence on consumption choices: A social network representation. *Heliyon, 9*(6).

Shell. (2022, December 16). *Shell completes acquisition of solar solutions provider Daystar Power Group.* Shell. Retrieved on February 02, 2025 from https://www.shell.com/what-we-do/renewable-power/renewable-power-news-releases/shell-completes-acquisition-of-solar-solutions-provider-daystar-power-group.html

Shen, W., & Power, M. (2017). Africa and the export of China's clean energy revolution. *Third World Quarterly, 38*(3), 678–697.

Shivji, I. (2009). *Accumulation in an African periphery: A theoretical framework.* Mkuki na Nyota Publishers.

Simmet, H. R. (2018). "Lighting a dark continent": Imaginaries of energy transition in Senegal. *Energy Research & Social Science, 40*, 71–81.

Smith, A., Kern, F., Raven, R., & Veerhes, B. (2014). Spaces for sustainable innovation: Solar photovoltaic electricity in the UK. *Technological Forecasting and Social Change, 81*, 115–130.

Song, J., & Ireri, B. (2024, August 30). *How China and Africa can better collaborate to close Sub Saharan Africa's energy access gap.* World Resources Institute. Retrieved on August 15, 2025 from https://www.wri.org/update/africa-solar-belt-implementation-focac-2024

Stock, R., Nyantakyi-Frimpong, H., Antwi-Agyei, P., & Yeleliere, E. (2023). Volta photovoltaics: Ruptures in resource access as gendered injustices for solar energy in Ghana. *Energy Research & Social Science, 103*, 103222.

Sun, S. (2013). *Funding breakthrough technology, case summary: Photovoltaics.* University of Cambridge. Retrieved on January 30, 2025 from https://www.jbs.cam.ac.uk/wp-content/uploads/2023/06/cbr-photovoltaics-case-report.pdf

Sustainable Energy for All. (2023). *SEforALL CEO, President of Kenya, country partners unite around key energy transition efforts.* Retrieved on February 14, 2025 from https://www.seforall.org/news/seforall-ceo-president-of-kenya-country-partners-unite-around-key-energy-transition-efforts

Sustainable Energy for All. (2024). *Manufacturing in Africa showcase: Auxano Solar*. Retrieved on March 3, 2025 from https://www.seforall.org/system/files/2024-08/remi-casestudy-auxano-solar-002.pdf

Taylor, I. (2016). Dependency redux: Why Africa is not rising. *Review of African Political Economy, 43*(147), 8–25.

The Economist. (2024, June 18). Private firms are driving a revolution in solar power in Africa. *The Economist*. Retrieved on February 19, 2025 from https://www.economist.com/middle-east-and-africa/2024/06/18/private-firms-are-driving-a-revolution-in-solar-power-in-africa

The Voice of Africa. (2025, April 21). *Are Trump's tariffs a wake-up call for Africa?* The Voice of Africa. Retrieved on June 18, 2025 from https://thevoiceofafrica.com/2025/04/21/are-trumps-tariffs-a-wake-up-call-for-africa/

Thompson, C. (2018). Philanthrocapitalism: Rendering the public domain obsolete? *Third World Quarterly, 39*(1), 51–67.

Trompette, P., Etienne, E., & Francius, R. (2022). At the margins of the grid: The politics of off-grid electrification in Senegal. In N. Ojong (Ed.), *Off-grid solar electrification in Africa: A critical perspective* (pp. 65–110). Springer.

Tsikata, D. (2006). *Living in the shadow of the large dams: Long term responses of downstream and lakeside communities of Ghana's Volta River Project*. Brill.

United Nations Conference on Trade and Development [UNCTAD]. (2022). *Economic development in Africa Report 2022: Rethinking the foundations of export diversification in Africa—The catalytic role of business and financial services*. United Nations. Retrieved on February 25, 2025 from https://unctad.org/system/files/official-document/aldcafrica2022_en.pdf

United Nations Conference on Trade and Development [UNCTAD]. (2023a). *The state of commodity dependence 2012*. United Nations Conference on Trade and Development.

United Nations Conference on Trade and Development [UNCTAD]. (2023b). *Technology and innovation report 2023: Opening green windows: Technological opportunities for a low-carbon world*. United Nations. https://unctad.org/system/files/official-document/tir2023_en.pdf

United Nations Development Programme [UNDP]. (2011). *Towards human resilience: Sustaining MDG progress in an age of economic uncertainty*. United Nations Development Programme. Retrieved on February 26, 2025 from https://www.undp.org/sites/g/files/zskgke326/files/publications/Towards_SustainingMDG_Web1005.pdf

United States Agency for International Development [USAID]. (2017). *Scaling access to energy in Africa*. USAID. Retrieved on February 26, 2025 from https://pdf.usaid.gov/pdf_docs/PA00WJN1.pdf

Velasco, A. (2002). Dependency theory. *Foreign Policy, 133*, 44.
Vietor, R. (1984). *Energy policy in America since 1945*. Cambridge University Press.
Wald, M. (1989, August 3). ARCO to sell Siemens its solar energy unit. *The New York Times.*
Wallerstein, I. (1974). *The modern world-system I: Capitalist agriculture and the origins of the European world-economy in the sixteenth century.* Academic Press.
Wallerstein, I. (2004). *World-systems analysis: An introduction.* Duke University Press.
Wang, X., Yu, C., & Wei, Y. (2012). Social media peer communication and impacts on purchase intentions: A consumer socialization framework. *Journal of Interactive Marketing, 26*(4), 198–208.
Weber, H. (2004). The "new economy" and social risk: Banking on the poor? *Review of International Political Economy, 11*(2), 356–386.
Wong, S. (2010). *Elite capture or capture elites? Lessons from the 'counter-elite' and 'co-opt-elite' approaches in Bangladesh and Ghana* (No. 2010/82). WIDER working paper.
Wood Mackenzie. (2019). *Strategic investments in off-grid energy access: Scaling the utility of the future at the last mile.* Power & Renewables Research Report in partnership with Energy 4 Impact.
World Bank. (2020a). *World Bank project to boost household access to affordable energy*. World Bank. Retrieved on February 07, 2025 from https://www.worldbank.org/en/news/press-release/2020/09/17/world-bank-project-to-boost-household-access-to-affordable-energy#:~:text=Rwanda%20EAQIP%20aims%20to%20improve,off%2Dgrid%20solar%20home%20systems
World Bank. (2020b). *Lighting up Africa: Bringing renewable, off-grid energy to communities.* World Bank. Retrieved on February 07, 2025 from https://www.worldbank.org/en/news/feature/2020/08/13/lighting-up-africa-bringing-renewable-off-grid-energy-to-communities
World Bank. (2024). *Off-grid solar could provide first-time electricity access to almost 400 million people globally by 2030*. Retrieved on February 07, 2025 from https://www.worldbank.org/en/news/press-release/2024/10/08/off-grid-solar-could-provide-first-time-electricity-access-to-almost-400-million-people-globally-by-2030
World Bank. (2025). *Mission 300 is powering Africa.* World Bank Group. Retrieved on February 07, 2025 from https://www.worldbank.org/en/programs/energizing-africa/overview

4

Solar Finance and Everyday Extraction

4.1 Introduction

Solar electrification in Africa is more than an account of technology diffusing into 'last mile' markets or finance scaling across borders. It is, above all, about insertion and depletion. I show how households are inserted into repayment ecologies that discipline everyday survival into collateralized flows, while depletion registers in the bodily, ecological, and temporal costs that stabilize those flows. What is often celebrated as empowerment (Khasru & Rafee, 2021) is in effect dependence: insertion into fragile debt circuits that convert missed payments into darkness, and depletion of marginalized people's time, food budgets, and reputational standing to keep those circuits intact.

To grasp how these intertwined processes take shape, I approach financialization as the very mechanism through which insertion becomes legible and depletion becomes normalized. At first glance, these practices might be read as evidence of 'portfolio management' by financially savvy households, a narrative trumpeted by financial inclusion advocates (see Collins et al., 2009), but that view obscures more than it reveals. These improvisations are not minor acts of household budgeting; they are constitutive practices that continually reshape financialization

N. Ojong, *Solar Power Capitalism*, Energy, Climate and the Environment,
https://doi.org/10.1007/978-3-032-13707-4_4

from within. By stretching repayment schedules, repurposing savings groups, or negotiating arrears, households reconfigure the infrastructures of pay-as-you-go (PAYG) solar home system (SHS) repayment on which securitization depends. What seems as survival is in fact the everyday labor that sustains and reshapes financial circuits, ensuring that global flows of repayment remain reliable even under conditions of local precarity.

Ethnographic work across different contexts reinforces this argument. As Kaushal (2025) shows in her study of microfinance in West Bengal, the very appearance of standardized finance depends on the affective and social-reproductive labor of women intermediaries who reconcile profit imperatives with the moral economies of clients. This resonates with the East African PAYG SHS landscape, where branch staff and local agents similarly translate abstract repayment schedules into enforceable trust.

To clarify the structure of this analysis, I organize the chapter into two movements. Movement I focuses on four intermediary conduits: PAYG platforms, local suppliers and installers, indigenous savings and credit institutions, and peer lenders, showing how each establishes an accumulation niche while outsourcing discipline to digital tools or social obligation.[1] Movement II reveals how kilowatts become data, data become collateral, and collateral underwrite securities marketed thousands of miles away. This is not only financialization; it is the machinery of insertion and depletion. Without insertion, no data stream exists; without depletion, no circuit stabilizes. This dual architecture sets the stage for the next section. To understand how insertion and depletion are lived, I turn to the intermediary conduits through which households access solar energy. These conduits are infrastructures that bind everyday life to global finance.

[1] In this book, I use Indigenous financial institutions and indigenous savings and credit institutions interchangeably.

4.2 Intermediary Conduits

These conduits must be read as infrastructures of insertion and depletion: every mechanism that delivers light also incorporates households into fragile circuits, while draining the liquidity and trust that keep repayment possible. Financialization and fintech are not the phenomenon themselves but the apparatus through which insertion is operationalized and depletion rendered invisible.

4.2.1 PAYG Platforms: Insertion Through Digital Leverage, Depletion Through Automated Darkness

PAYG platforms illustrate how insertion and depletion are engineered through digital infrastructures.[2] Households are inserted into repayment circuits where every mobile transfer and kilowatt becomes collateral, while depletion registers in the social trust, liquidity, and gendered labor stretched to keep access to light intact. Building on these dynamics, PAYG solar firms capture not only kilowatt-hours but also behavioral metadata, converting them into revenue streams that can be securitized at scale.

Seen from this perspective, PAYG operators epitomize the platform political economy of fintech: reintermediating financial flows through digital infrastructures, consolidating market power in a few firms, and capitalizing speculative expectations through venture investment (Langley & Leyshon, 2021). Leveraging these platform dynamics, their SIM-enabled lockout turns late payment into an instantaneous asset-protection tool, allowing investors to hedge default risk from afar.

---

[2] For example, upOwa, a Franco-Cameroon off-grid solar enterprise supported by the Renewable Energy Performance Platform (REPP), reveals how PAYG operators insert low-income households into data-driven repayment circuits. Through its rent-to-own model and mobile-money infrastructure, upOwa extends electrification to rural areas while embedding energy access within continuous financial obligations. See Renewable Energy Performance Platform (REPP), upOwa Solar Home Systems project, https://repp.energy/project/upowa-solar-home-systems/.

This logic did not emerge in a vacuum. The rise of PAYG solar power initiatives in East Africa cannot be separated from what Knight and Wójcik (2020: 1490) term the 'FinTech fever,' catalyzed by M-Pesa's 2007 launch and the subsequent proliferation of mobile-payment infrastructures that reshaped finance's spatial reach into rural and peri-urban areas.[3] In practice, once a household falls behind, the device is remotely disabled: no light, no TV, no phone charging. Everyday darkness therefore becomes collateral, since the withdrawal of electricity itself pressures families to repay; continued access depends on regular mobile-money transfers. What once required loan officers, threats of repossession, or costly face-to-face collection is now automated through a SIM chip and digital network. This automation lowers enforcement costs and produces predictable repayment streams that investors classify as low-risk assets.

These shifts in enforcement parallel a deeper transformation in how value is generated. They mirror the platformization of value (Roitman et al., 2025), where data streams themselves—kilowatt-hours, payment metadata, user behavior—become assets that can be priced, securitized, and circulated beyond the original site of production.

But beyond investor portfolios, such practices are embedded in broader development agendas. They are the outcome of a fintech–philanthropy–development complex that tracks and monetizes the digital footprints of the poor, converting households into legible financial subjects whose everyday transactions are profiled and securitized (Gabor & Brooks, 2017).

While these global circuits frame the macro-logic of solar finance, the household scale reveals its human cost. At this level, the discipline of PAYG systems often compels users to borrow from neighbors or indigenous credit institutions to restore supply, shifting the burden of asset protection from financiers onto local social relations. PAYG platforms are hence not neutral providers of energy access but mechanisms of insertion and depletion. They insert households into datafied circuits of obligation, where every light bulb switched on is collateralized, while

[3] For a rigorous critique of the celebrated poverty-reduction claims attached to M-Pesa and the wider fintech-for-development agenda, see Bateman et al. (2019).

depleting liquidity, social trust, and especially women's labor as families juggle flour-or-light trade-offs to maintain access.

The lived consequences of these circuits become clearer when examined through the pricing and payment modalities of off-grid solar products themselves. The high upfront cost of most off-grid solar (OGS) products renders them unaffordable for a significant segment of the population (Bhattacharyya & Palit, 2016; Rolffs et al., 2015). It is at this juncture that the PAYG models, which exemplify the convergence of solar technologies with structured financial instruments, play a vital role.

As noted in Chapter 1, PAYG models leverage modern information technology and enhanced mobile connectivity to facilitate flexible online repayments via the mobile phone, thereby increasing SHSs affordability (Ockwell et al., 2019; Ojong et al., 2025; Yadav et al., 2019). In most PAYG contracts, clients are often presented with a rent-to-own option, where they begin with an initial down payment and subsequently make fixed monthly installments over a predetermined period, typically ranging from 18 to 36 months, after which ownership of the system is transferred to them. Alternatively, some providers offer a 'usage-based payment' plan, also known as the 'energy-as-a-service' approach, whereby clients prepay for the electricity they consume but never gain ownership of the system itself, maintaining access only as long as they continue making payments (Brown et al., 2022; Riedke & Adelmann, 2022). In the communities covered in this book, PAYG SHS companies offer the rent-to-own plan. While mobile money is the primary payment method, some companies also accept cash or provide scratch card options. PAYG SHS companies in Africa such as Sun-King and M-Kopa require an initial down payment before installing the system, after which customers can decide how much to pay at regular intervals (e.g., weekly, biweekly, etc.) to access energy services, though some companies set a minimum payment requirement, as part of the rent-to-own model, as described by some end-users in Kenya and Tanzania:

> I paid a deposit of KES1800 [US$14] and pay KES40 [US$0.31] each day (DG, Kenya, 2023)

> It [payment] was not per day but weekly. The KES930 [US$7.2] was inclusive of KES23 [US$0.18] transaction fee (AE, Kenya, 2023).
> I was paying little-by-little through buying vouchers in order to get light (TS, Tanzania, 2022).[4]

These accounts illustrate that what looks like flexibility in repayment is, in effect, the mechanism by which insertion and depletion unfold: insertion into circuits that couple light to liquidity, and depletion of food budgets and social trust to keep those circuits intact.

While PAYG models substantially reduce the financial barriers associated with obtaining SHSs, the experience of end-users is shaped by nuanced payment structures that intertwine accessibility with ongoing obligations. For example, one of the Kenyan SHS user cited earlier highlighted the requirement of initial deposits (such as KES1800 [US$14]), followed by daily or weekly installments that can accrue additional fees, thereby illustrating the potential for staggered payments to become an everyday expense rather than a single, upfront investment. Similarly, the Tanzanian account of purchasing 'vouchers' to unlock incremental usage shows how flexible repayment modalities can be advantageous but also embed end-users in a system where energy access hinges on steady, smaller transactions. These models undoubtedly broaden the SHS customer base by allowing low-income households to bypass prohibitive lump-sum expenditures; however, they simultaneously create new forms of financial dependency by merging energy consumption with consistent micro-payments.

Equally revealing are users' remarks about the daily negotiations required to meet these mini payments. One Kenyan participant noted that 'with KES930 [US$7.2] and the cost of living is also currently high, you contemplate whether to buy flour or pay for the solar [SHS]. Sometimes it's difficult' (PT, Kenya, 2023). This dilemma is not simply a personal budgeting problem; it illustrates how solar energy access is governed by infrastructures that couple repayment to subsistence, thereby linking household consumption choices directly to financial architectures.

---

[4] The exchange rate used throughout this book is US$1 = KES128.

Comparable trade-offs have been documented in very different settings. In the United Kingdom, Burlinson et al. (2022) show how prepaid electricity meters generate a 'heat-or-eat dilemma' in low-income households, forcing families to choose between maintaining warmth and buying food. The Kenyan participant's reflection that she must weigh paying for solar energy against buying flour resonates with this dilemma. While the specific socio-economic landscapes differ, both cases stress how prepaid energy infrastructures impose a discipline of rationing that ties energy access directly to the capacity to forgo other necessities. These trade-offs demonstrate that what appears as an individual budgeting struggle is in fact structurally engineered by prepaid infrastructures. Each choice at the kitchen table is simultaneously a data point in repayment streams that are bundled and sold abroad, illustrating how fundamental dilemmas of food and light sustain the securitized architectures of global finance.

Set alongside the opening claim that household improvisation co-produces finance, these global resonances, whether in Kenya's flour-or-light trade-offs or Britain's heat-or-eat choices, show that the discipline of prepaid energy is not restricted locally; it is a structural feature of contemporary financialization.

Such experiences highlight how an ostensibly convenient PAYG model can become burdensome under tight financial conditions, compelling households to juggle multiple obligations and, at times, borrow from one's social network to avoid disconnection.

These micro-repayment cycles, while advertised as 'flexible,' effectively link energy access to consistent financial commitments. One Tanzanian mother stated she 'never prefers to be in debt' (ZW, Tanzania, 2022) yet feels compelled to pay weekly to maintain her household's lights. Her reluctance reveals how the affective burden of debt is folded into technical payment systems, showing how individual dispositions toward borrowing are mobilized to stabilize repayment flows attractive to financiers. In other words, the practical benefit of incremental financing, no large upfront cost, comes with heightened pressure to budget around solar energy installments.

Consequently, PAYG can transform basic energy consumption into a continuous repayment regimen (Baker, 2023). Everyday life is punctuated by countdowns to the next mobile transaction, ultimately embedding solar power access within a web of financial obligations. From this lens, even intimate feelings about borrowing are folded into repayment infrastructures. Reluctance and aspiration alike become instruments of financial discipline, stabilizing cash flows that investors thousands of miles away regard as predictable assets. Household affect, the anxiety of debt or the pride of repayment, thus becomes part of the circuitry of accumulation, stabilizing financial flows that investors classify as reliable assets.

The circulation of loans among kin, indigenous financial institutions, and neighbors could also be mistaken for resilience. In reality, these short-term fixes expose the fragility of household finances under PAYG solar power. Missing a payment trigger not only darkness but also reputational damage, transforming minor income shocks into cascading liabilities. Rather than deepening resilience, such arrangements redistribute the costs of financial risk downward, intensifying dependence on volatile incomes and informal debts. This deepening dependence becomes especially visible in users' accounts about upgrades and repeat purchases, moments that appear aspirational but extend the cycle of indebtedness.

Once people are in the web, it becomes challenging to escape from the financial circuit as participants in Kenya explained:

> I took the first solar home system for lighting, which I finished paying for. They told me that when I finished paying, they could bring me a TV without a deposit. So, I just paid money for the first week for the TV (GT, Kenya, 2023).
>
> They [PAYG SHS company] called us saying that they could see our account was clean, we had settled the debt we had. They offered to give us a TV but that required a bigger solar home system. We re-registered with them again, but they did not even ask for a deposit. For this one [bigger SHS], we pay KES50 [US$0.38] per day. This bigger one [SHS] is still powering all the appliances (BT, Kenya, 2023)
>
> When I was about to complete the payment for the solar home system, with only KES600 [US$4.6] remaining, I received a call informing me

> to go and collect the TV. I went there and was given the TV along with two batteries combined as one unit. These batteries only function when connected together. They were configured to power both the TV and the lighting system, which I had already fully paid for. The system was set up in such a way that any further payments I made would go toward the TV, since the lighting system was already settled. As a result, both the TV and the lights were operational. However, if I failed to make payments, neither would work. I persisted until I fully paid for the TV. Once I completed the payment, they called me again and gave me 24 iron sheets (ZU, Kenya, 2023).

These upgrades are not incidental rewards but carefully structured inducements that convert household aspirations into longer debt cycles, ensuring that even moments of apparent upward mobility generate fresh collateral for global investors. What looks locally like aspiration, expanding from a single light to a television or iron sheets, feeds directly into accumulation logics that prize continuous revenue streams. Each upgrade is not simply a new appliance but a mechanism for deepening households' entanglement in securitized debt cycles.

Beyond individual upgrades, the durability of this debt relationship rests on a deeper logic: PAYG companies systematically track and formalize credit histories, transforming household repayments into a metric of risk and worth. This credit history then enables them to differentiate between creditworthy and uncreditworthy clients, with more credit, in the form of solar-powered products, being extended to those in the former group. Credit histories do not just reflect prior behavior; they formalize and recast social relationships as metrics of worth, enabling both surveillance and selective inclusion into circuits of capital. Then, based on clients' credit histories, PAYG solar companies appear to cultivate a de facto revolving line of credit that links precarious consumers of OGS power into broader circuits of financial capital. The ability to build credit histories sets PAYG platforms as cost-efficient intermediaries in the process of financial accumulation.[5]

[5] A Bloomberg investigation reported that while utility-scale solar is the cheapest new energy globally, the micro-debt structure of PAYG schemes has rendered household solar power prohibitively expensive for the very populations it targets (Kocieniewski & Finch, 2022).

More recent work reframes this dynamic through the lens of risk. Instead of only seeing debt-for-access as the extension of credit to the unbanked, it is equally a mechanism for redistributing volatility (Mader et al., 2025). PAYG schemes commodify uncertainty itself, shifting the risks of currency depreciation, fluctuating incomes, or unexpected expenditures onto households (Mader et al., 2025), while insulating investors through securitized repayment streams and remote lockout. This perspective situates debt-for-access not only as inclusion but as a form of risk transfer.

This dynamic also directly intervenes in long-standing debates on microfinance and financial inclusion (Roitman, 2025). Some researchers have long argued that linking repayment histories to new credit opportunities fosters empowerment and economic mobility (Armendáriz & Morduch, 2010; Morduch, 1999). However, critics show that such platforms connect low-income households to new circuits of indebtedness, risk, and discipline, often entrenching rather than alleviating precarity (Bateman, 2010; Mader et al., 2025; Soederberg, 2013; Weber, 2014). PAYG solar finance both draws on and reconfigures these logics: by transforming energy access into a credit-scoring mechanism, it places households at the intersection of infrastructure provision and financial accumulation.

In Kenya and Tanzania, households described taking loans from indigenous savings and credit institutions to keep up with weekly PAYG payments, effectively layering one form of micro-debt onto another. Here, PAYG does not only include the poor in finance, but actively reworks the field of financial inclusion by turning everyday energy consumption into an index of creditworthiness. To understand the affective pull of this system, it is necessary to examine how desire and consumer aspiration are mobilized as instruments of control.

PAYG solar companies strategically couple consumer desire for new devices with a payment mechanism that entrenches financial obligations. On the one hand, acquiring a TV appears empowering: an expansion of energy-driven possibilities beyond basic lighting. However, this bundling also deepens dependency on the company's payment structure. Missing installments not only interrupts the enjoyment of the TV; it can also disable devices already paid for. In this way, solar technology becomes a tool for financial capital expansion, where the promise of energy autonomy, framed as liberation from kerosene, candles, or unreliable grids, is instead structured through debt obligations that bind households to corporate control. Autonomy is conditional and easily withdrawn, as non-payment translates immediately into disconnection. These acts of borrowing and upgrading, though entangling, are often pursued as forms of social aspiration and householding, where financial obligations intertwine with the relational work of sustaining kin and future hopes (see also Zaloom and James, 2023). This conditionality reaches its sharpest expression in the ambiguous meaning of 'ownership' itself.

PAYG repayment models do more than grant households access to lighting or TVs: they blur the line between possessing a device and perpetually owing for its use (Riedke & Adelmann, 2022). A Kenyan household that had finished paying for one Azuri SHS kit found it reliable and entered a new two-year contract for a TV upgrade. When installments fell into arrears, the company remotely disabled the newer system, leaving the family without the energy service they believed they already owned. Such cases reveal that 'ownership' is precarious: the solar panel may sit on the roof (Fig. 4.1), but functional control is ceded to the firm until payment is complete, and the household does not contract a new debt. The question of who really owns the system, the household or the company, remains nuanced.

Fig. 4.1 Solar panel on a roof, Kenya (Author's photo taken during fieldwork)

These blurred arrangements also lay the groundwork for the next cycle: the conversion of repayment history into a marker of creditworthiness. As mentioned earlier, most PAYG firms often offer additional devices the moment a prior debt is settled, turning energy access into a visible symbol of social mobility (see also Cahn, 2008; Riedke & Adelmann, 2022). The rooftop solar panel signals aspiration, while timely repayment generates the data that qualifies households for the next loan or upgrade. Creditworthiness is both a personal achievement and a criterion for further evaluation, compelling families to shoulder fresh debts in the hope of advancement.

I frame this as *recursive creditworthiness*: a self-reinforcing cycle in which repayment sustains access while simultaneously producing the evidence required for future borrowing. The household is drawn into an ongoing cycle of indebtedness, where current inclusion becomes the condition for deeper entanglement over time. In effect, a household's ability to keep paying not only preserves its current access to light but also anchors the value of financial products marketed abroad. Recursive creditworthiness thus operates at two levels: for the family seeking the next upgrade and for investors seeking reliable cash flows from aggregated repayment data.

The automatic lockout feature that suspends energy services when users fall behind on payments operates as a potent mechanism of control within the PAYG solar sector in Africa. In fact, that feature makes PAYG models very attractive financial intermediaries. This is precisely because in addition to reducing transaction cost of payment collection, it decreases investment risk (Tongsopit et al., 2016; Zhang, 2014). Clients of PAYG solar companies highlighted the automatic lockout feature of their SHSs:

> I was disconnected seven times (DF, Kenya, 2023).
> When I am used to something [solar electrification] and then it becomes unavailable, I feel very bad because, during that time, I cannot even charge the radio, nor can I charge the torch (LU, Kenya, 2023)
> Solar Panda stopped functioning until I made the payment (TX, Kenya, 2023).

Here again, everyday experiences of disconnection—darkened evenings, strained dignity—are not just local misfortunes. They are the very contradictions that allow investors to hedge risk, transforming household vulnerability into tradable securities. Darkness itself becomes a financial instrument: the ability to cut light instantly converts household vulnerability into a predictable, tradable form of security for investors.

These household contradictions do not only stabilize financial flows; they also demonstrate how enforcement has shifted from human interaction to digital systems. Platforms wield power not by owning things but by owning the rules that mediate access to things (Srnicek, 2017). The SIM-enabled control board at the heart of every PAYG SHS is hence more than a control mechanism; it is a governance device that automates discipline in ways microfinance loan officers once performed manually (see Kar, 2013). When a Kenyan mother finds the lights off at 7 p.m., a digital system, not a person, has rendered judgment on her creditworthiness. That tiny act of 'digital violence' reveals how platform capitalism exports fintech logics into rural and peri-urban energy landscapes, radically lowering enforcement costs while shifting moral responsibility for darkness onto the debtor.

Viewed in this light, the lockout embodies more than digital violence; it reveals a subtler infrastructural power. Infrastructural power is the quiet authority that stems from embedding rules into everyday technical systems (Petry, 2020; Pinzur, 2021; Roitman, 2025). Unlike coercive state power, it works by making people dependent on infrastructures that appear neutral or inevitable (Pinzur, 2021). In this case, the ability of firms to turn lights off remotely illustrates how control is exerted not through overt force but through dependence on a technical device. As recent work on financial platforms shows, infrastructural power lies in the capacity of firms to embed themselves in routines and dependencies, making their operations indispensable to survival (Roitman, 2025). The lockout function of PAYG SHSs exemplifies how private firms translate financial defaults into immediate interruptions to the daily organization of household life.

The social meaning of these interruptions becomes visible in communities where light itself functions as a symbol of respectability. The abrupt return to kerosene lamps, candles, or total darkness is more than an inconvenience; it becomes a visible sign of financial distress that undermines one's social standing. In communities where material goods are deeply entangled with recognition, light signals achievement while its absence suggests decline. Returning to the dull glow of a kerosene lamp can register as a symbolic demotion, a rank signal, marking the fragile boundary between those seen as 'successful' and those slipping behind.

What defines a class is not only material position but also how it is seen and acknowledged (Bourdieu, 1984). In this context, dim light becomes a barometer of worth. Neighbors notice, rumors circulate, and households find themselves doubly powerless: cut off from electricity and from the performative currency of respectability. These micro-defaults puncture the image of self-mastery projected by constant illumination and expose the thin layer of autonomy sustained under neoliberal precarity. The recurring rhythm of light and dark in some PAYG SHS households thus embodies a wider convergence of aspiration, social anxiety, and the quiet violence of market failure: threads that ripple into the bodily and ecological tolls discussed in Chapter 5.

PAYG firms are not the only intermediaries translating household desire into debt. At the community level, local suppliers and installers also serve as crucial sites of mediation where insertion and depletion take shape through incremental sales, technical know-how, and the diversion of scarce incomes.

4.2.2 Local SHS Suppliers and Installers: Insertion Through Sales, Depletion Through Scarce Incomes

Local SHS suppliers, a collective term for shop owners and agents, anchor solar energy markets at the community level, but they too operate as mechanisms of insertion and depletion. They insert households into long-term repayment obligations through incremental sales and technical services, while depletion surfaces in the diversion of already constrained household income, harvest proceeds, or reputational capital to maintain access. These actors embody the everyday intermediation that makes global circuits of finance legible at the local level.

The local shop owners acquire these systems from wholesalers to sell them in their communities. But they are not alone. Alongside shopkeepers, local agents, who are not employees of OGS companies, earn commissions from the sale of SHSs. In addition, freelance solar PV installers, i.e., self-employed individuals, provide installation services. Most SHS companies are responsible for installing the SHSs after purchase. But a significant percentage of the population does not purchase their solar PV systems from these companies. The presence of solar PV installers becomes especially critical where households acquire systems outside corporate distribution channels. In such cases, installation becomes a negotiated act: sometimes do-it-yourself, sometimes paid. SHS users' accounts vividly reveal this mix of autonomy and expenditure.

As one participant in Cameroon put it: 'I did the installation myself, no need to pay anybody' (CR, Cameroon, 2024). Another participant in Cameroon remarked, 'I paid a technician XAF5,000 [US$9] for installation' (JI, Cameroon, 2024).[6] These contrasting accounts point to a vital but overlooked dimension of solar power diffusion: the everyday financial transactions and practical know-how that make technology functional. These actors, too, embody circuits of insertion and depletion. Their incremental sales and installation services insert households into long-term repayment obligations, even outside formal PAYG contracts, while depleting precarious household incomes, harvest revenues, and reputational capital that families stake in order to maintain access.

In Malawi, similarly diverse interactions between local suppliers and solar PV installers are evident. One user noted a direct purchase of SHS components following a successful agricultural sale, stressing self-reliance: 'After selling my onions, I went to the local shop and purchased the solar panel and battery. I installed it myself' (UH, Malawi, 2022). Conversely, another user reflected on the recurring dependence on a local shop for battery replacements, illustrating the cyclical nature of local market relationships: 'Usually, I have to go back to the same shop

[6] The exchange rate used throughout this book is US$1 = XAF$558.

where I bought it [battery] to get another one' (JI, Malawi, 2022). The reliance on local shopkeepers does not only position these individuals as commercial intermediaries but reveals their role in perpetuating cycles of dependency, inadvertently reinforcing ongoing financial commitments for end-users.

Another account underscores this duality by emphasizing the indispensability of skilled labor. 'The shop owner sold me the parts separately,' a user recounted, 'then I paid a technician MWK 25,000 [US $14.4] to install the system at home' (SX, Malawi, 2022). Here, the shop owner and solar PV installer play two intimately linked intermediary roles. The former enables people to become owners of SHSs while the latter enables the new owners to put their devices into use. As I show later, their intervention enables financial capitalists to get returns on their investment.

Comparable exchanges appear in Tanzania, where neighborhood trust and small favors lubricate solar transactions. In that country, relationships of proximity often mediate payment flexibility. One user explained, 'I paid a technician, who is my neighbor, to install the solar panel on the roof. It cost TZS15,000 [US$6.2] to install it. I paid him TZS10,000 [US$4.1] upon completing the work and finished paying the balance the next day' (OV, Tanzania, 2022). The solar PV installer's neighborly status facilitates flexible payment terms, subtly intertwining economic transaction and social trust. In other words, in this local context, social trust greases the wheels of solar power capitalism. Along similar lines, another SHS user who lacked the requisite expertise straightforwardly acknowledged, 'Because no one at home could install the solar panel on the roof, we paid a technician TZS25,000 [US$10.2] to install it' (NX, Tanzania, 2022). This acknowledgment further situates local solar PV installers as indispensable actors within the broader architecture of solar energy capitalism, reinforcing their intermediary capacity between technological access, financial commitments, and household energy security.

Across Kenya and Tanzania, local SHS suppliers and installers reveal nuanced intermediary roles within a complex landscape that merges financial dependency with social connectivity. A user in Tanzania explained, 'Most people around here use Sun-King. They pay TZS20,000 [US$8.2] for the deposit, and then the local agent makes sure we know how to operate it. If something breaks, we either call the agent or look for a technician in the next village' (RJ, Tanzania, 2022). Once again, this shows how SHS installers, operating beyond the bounds of formal corporate affiliations, maintain the continuity of solar services, becoming indispensable intermediaries who step in precisely when formal channels become unreliable or inaccessible. This narrative is echoed by Kenyan experiences, where delays in corporate responsiveness shift responsibility to local agents, as one user explained: 'When the SHS stopped working, we gave it to the agent from whom we bought it so that he could take it back to the solar company for repair. When we did not hear back from the agent for a while, we pressurized him, and he went and collected it [SHS] from the solar company' (HF, Kenya, 2023). These examples reveal how informal actors backstop formal supply chains, absorbing reputational risk and financial strain that companies deflect.

Additionally, the practice of incremental payments facilitated by local suppliers further highlights the intricate balance between community relationships, financial flexibility, and solar power access. This flexibility is succinctly captured by an SHS user in Tanzania who noted: 'We paid little-by-little. The shop owner allowed us to pay that way, and when we completed payment, we took our SHS' (BX, Tanzania, 2022). Such arrangements show the role of local suppliers as crucial financial intermediaries who render these devices more affordable. Unlike PAYG SHS contracts, these payment-by-installment arrangements mean that households take full ownership of the device once the final installment is paid to the shop owner, with no risk of remote disconnection. But transnational financial capitalists still get returns on their investment indirectly: local suppliers purchase their stock from wholesalers and importers linked to global investors, so every unit sold through these community-based payment plans still feeds the broader circuits of capital accumulation. Transnational financial capitalists get returns on their investment through such location-specific initiatives that render SHSs affordable. What appears as a purely local arrangement, a shopkeeper extending flexible repayment terms, feeds into global circuits of accumulation. The shop owner's ability to sell stock rests on supply chains financed by importers and wholesalers tied to transnational capitalists, meaning that incremental local payments help secure hard-currency returns upstream.

Malawi adds another dimension: the coexistence of outright cash purchases and installment arrangements. Some households, often after a good harvest, reported buying solar panels and batteries in full, thereby achieving immediate ownership. Others relied on installment plans through local shops, paying piece-by-piece until the system was theirs. While the former appears insulated from PAYG's cycles of remote disconnection, they were not free of precarity: a single failed harvest could mean no funds for repairs or replacements, leading families back into cycles of informal borrowing. Installment buyers, meanwhile, spoke of constant pressure to keep up payments, even when harvests were poor. This contrast highlights how solar energy capitalism operates along multiple pathways, cash sacrifice, incremental installment purchase, and PAYG debt, all of which redistribute risk downward, albeit in different temporal rhythms.

Equally significant is how these flexible payment arrangements reshape the broader contours of local power and finance. By accepting incremental payments, shopkeepers provide a local banking function for solar firms. In doing so, they commodify trust, selling not only SHSs but also their reputational standing within the community. By allowing incremental payments, these shop owners leverage community-based trust and social capital to mitigate the risks typically associated with lending to this segment of the population.[7] In essence, they absorb some of the financial uncertainty on behalf of customers who might otherwise be excluded from OGS electrification. At the same time, their role as gatekeepers of solar power access can reinforce new forms of dependency and obligation, albeit in ways that reflect established social norms, reciprocal expectations, and collective notions of mutual aid. People engaged in such debt relationships with local shop owners must demonstrate reliability to the shop owner, ensuring that the latter's own risk, such as non-payment, remains tolerable. This dynamic ties energy access to personal reputation and reinforces a debt-like relationship, wherein one

---

[7] On the dynamics of trust and social capital, see Ojong and Simba (2019), Ojong (2018).

party feels compelled to behave in ways that does not jeopardize future access.

Because of the favorable payment arrangements, the buyer of the SHS may feel morally compelled to reciprocate via recommendations to friends and neighbors, or small favors in other community settings.

Such transactions reveal that financial capital, whether circulating through global markets or localized economies, is never detached from power and morality. Local suppliers and installers transform community trust into a mechanism of accumulation, ensuring that every kerosene lamp replaced by solar-powered electric lighting also activates new circuits of obligation. The next intermediary, indigenous financial institutions, extends these dynamics of trust and reciprocity, repackaging them into collective systems that convert solidarity into collateral.

4.2.3 Indigenous Financial Institutions: Insertion Through Collective Collateral, Depletion of Reciprocity

Indigenous financial institutions, known in the scholarly literature as Rotating Savings and Credit Associations (RoSCAs) and Accumulating Saving and Credit Associations (ASCAs), convert reciprocity into liquidity, funneling transnational capital into local communities while earning interest for the collective. They illustrate how insertion and depletion unfold through collective financial practices. Households are inserted into repayment circuits by pooling contributions that lenders treat as reliable collateral, while depletion emerges in the erosion of reciprocity and solidarity that once buffered families from volatility.

These institutions thus function both as infrastructures of survival and as mechanisms that sustain securitization downstream. Flexibility lowers default rates and financial stress but yokes communal solidarity to investor returns. These indigenous financial institutions have long played a defining role in financing small but crucial expenditures in African economies (Ardener & Burman, 1995; Guyer, 2004; James, 2012; Ojong & Obeng-Odoom, 2017; Ojong, 2019a; Shipton, 2010). Beyond their better-known function of supplementing household cash flows, these indigenous savings and credit institutions serve as financial intermediaries between transnational financial capitalists and consumers of OGS technologies. By connecting buyers to financial resources, they enable transnational financial capitalists to penetrate deeper into local markets. Also, they reveal how deeply solar power capitalism has woven itself into everyday life in rural, peri-urban, and urban communities across Africa. A case from Malawi captures both the patience and vulnerability embedded in these arrangements.

'To purchase the SHS, I was saving money every year after selling my farm produce. It took me three years to save the funds needed to purchase the SHS. I saved the money through a village savings group' (MD, Malawi, 2022). This excerpt brings to the fore a protracted but dedicated effort toward building sufficient funds through a locally-embedded financial mechanism. This seemingly slow, patient act of saving within an indigenous financial institution is simultaneously a contribution to the securitized stability of solar firms, which depend on such collective liquidity to reassure investors. So, indigenous savings and credit institutions act as infrastructures of insertion and depletion: they insert households into circuits where communal savings become collateral.

Similar dynamics are visible across East Africa, where the same communal mechanisms finance deposits for PAYG systems. In Kenya, indigenous financial institutions became central to both household energy access and the securitized flows that underwrite the PAYG solar economy, as KF explained:

> I used KES7,000 [US$54], which I saved through a village savings group, as a deposit to get a solar TV. I am a member of four village savings groups. We save regularly and earn interest on our savings when people borrow money from the group. For example, you can save KES6,000 [US$46.4] in a year, and when the time comes to distribute the savings, you may find that your initial KES6,000 [US$46.4] has doubled, depending on how many people borrowed money from the group. At the end of the year, I had saved KES44,000 [US$340] with the group, and I used KES7,000 [US$54] to purchase a 32-inch solar TV (KF, Kenya, 2023).

KF's account brings into view how these community-based institutions function not only as savings pools but also as revolving credit systems that circulate liquidity through the community. Once more, insertion is clear: communal savings are transformed into collateral legible to distant financiers. Depletion follows in the slow erosion of reciprocity, as practices once rooted in solidarity become connected to repayment schedules. Hence, they draw their strength from the cycle of saving, lending, and repaying with interest, a cycle that benefits not just individuals but also the collective (see also Rutherford, 2000). Crucially, the liquidity generated in these indigenous financial institutions does not remain bounded by the community. It underwrites the repayment schedules of PAYG SHS contracts, which in turn anchor the securitized revenue streams marketed in distant financial hubs. Communal solidarity hence becomes collateral for transnational capital.

Instead of depending solely on external credit, often laden with high interest rates or stringent payback timelines (Beck & Peria, 2006; Demirgüç-Kunt & Klapper, 2012; Simba et al., 2024), these SHS users take advantage of a communal savings and credit system that reflects local norms of trust and reciprocity (Anderson & Baland, 2002; Mianda, 2020; Ojong, 2019b). In cases like the one mentioned earlier, indigenous savings and credit institutions serve as financial intermediaries by providing a platform for individuals to save small amounts until they accumulate the required lump sum to make a deposit for a rent-to-own SHS or to cover the upfront cost of purchasing an SHS.

In other cases, some individuals use these community-based financial institutions as lending avenues that facilitate quicker access to SHSs. An SHS user in Tanzania explained this succinctly: 'I borrowed money to buy the SHS from a savings group. I borrowed TZS50,000 [US$20.4] and repaid TZS60,000 [US$24.5]. The total amount paid back included TZS10,000 [US$4.1] as interest. The rules of the group require borrowers to pay back the money within three months' (TJ, Tanzania, 2022). Here, the local savings group charges interest on the loan, an arrangement that inevitably generates earnings for the group itself.

A Kenyan participant's account extends this logic, showing how indigenous institutions mediate between household need and investor confidence:

> I took a loan of KES3,000 [US$23.2] from a self-help savings group toward the purchase of a solar TV. I borrowed the money because it was the school holiday, and during this period, as you know, children like watching TV. So, we bought the solar TV for them to watch so that they wouldn't have to go to another person's house. Borrowers can repay the loan in full at once or reduce the total amount to be repaid by making partial payments. Since I could not repay the full amount at once, I gradually reduced the outstanding balance. (KV, Kenya, 2023)

In this example, the capacity to repay in installments adds flexibility, lowering the risk of default while preserving the group's liquidity through partial and consistent loan recovery. It is precisely because of core features such as flexibility, negotiability, convenience, and reliance on trust and reciprocity (Bouman & Hospes, 1994; Guerin et al., 2011; Hossein & Bonsu, 2023; James, 2012; Servet, 1995) that such indigenous financial institutions remain attractive and pervasive across African communities.[8] These core features also enable transnational capitalists to earn returns on their investment. For instance, people obtain a loan from such institutions to purchase an SHS because they know that the loan can be repaid in small amounts and that they can renegotiate repayment plans. As a result, people who would otherwise be unable to purchase an SHS become owners of these technologies and, therefore, gain access to electricity.[9] Seen in this light, indigenous financial institutions are not only moral economies of reciprocity. They also provide the liquidity that steadies securitized revenue streams. What looks like communal solidarity in Movement I anchors the confidence of investors in Movement II.

---

[8] The unique features of these indigenous financial institutions explain why they are equally present among the African diaspora (see, for example, Hossein, 2024; Mianda, 2020).

[9] Studies have shown that repayment flexibility lowers default rates (Battaglia et al., 2019) and reduces financial stress (Field et al., 2012).

The final intermediary, peer lending relocates the burden of repayment from groups to individuals, where dignity, trust, and reputation become the last reserves of collateral.

4.2.4 Peer Lenders: Insertion into Relational Debt, Depletion of Dignity and Trust

Peer lenders, (i.e., friends, neighbors, and family members), extend small loans that prevent immediate default. Specifically, peer lenders supply loans that enable people to acquire SHSs and stave off PAYG lockout. In communities across Africa, people often request loans from individuals such as friends, neighbors, and family members (Ojong, 2019a; Shipton, 2010). By 'peer lending,' I'm referring to loans by friends, neighbors, and family members. The difference between indigenous financial institutions and peer lenders is that in the former, the loan is provided by a group while in the latter, it is provided by an individual.

Across Africa, peer lenders in local communities embody a distinct mode of financial intermediation. Loan-giving among peers is historically embedded in community life, where reciprocity and obligation structure expectations (Ojong, 2019b; Shipton, 2010). In many communities, lower-status or younger individuals anticipate support from wealthier or better-connected peers, who in turn gain reputational capital and reciprocal claims. This moral economy of lending is well documented in classic studies of African exchange and agrarian life (Guyer, 2004; Scott, 1976), and its ongoing vitality is evident in social practice.

Participants in Malawi, for example, described borrowing from neighbors rather than formal lenders to avoid exploitative interest, while Tanzanian households emphasized the comfort of turning to relatives and neighbors for temporary loans, given the trust and flexibility they afford. From this vantage point, the role of peer lenders is not simply a stopgap for those unable to access commercial banks or microfinance institutions; it is a historically durable mode of social support that continues to shape how households navigate both everyday needs and new energy debts. These micro-level practices are not marginal; they are the very fibers that later stitch into larger circuits of value extraction. As I show in Movement II, the capacity of global investors to generate returns depends directly on such everyday intermediations. Peer lending similarly functions as a mechanism of insertion and depletion. These intimate loans insert households into relational circuits that prevent default in the short term, while depleting dignity, trust, and reputational standing as financial obligation bleeds into intimate social ties.

Such intermediation unfolds in intimate, community-based networks of trust and reciprocity. For example, one borrower in Malawi described relying on this informal mechanism to purchase an inverter: 'I wanted to buy an inverter after my inverter stopped working, and I did not have enough money, so I borrowed money from a friend. I borrowed MWK 6,000 (US$3.5). I repaid the same amount I borrowed, and I paid all the amount at once' (MV, Malawi, 2022). Such micro-transactions contribute to the growth of solar energy markets across the continent. Additionally, as noted by the SHS user in Malawi, this form of financial intermediation is attractive as the loan is interest-free, he was provided a MWK6,000 (US$3.5) and repaid the same amount. This is another difference between indigenous savings and credit institutions and peer lenders. Loans from friends, neighbors, and family members are often interest-free while most indigenous financial institutions charge interest on loans.

Despite the relative accessibility of these peer networks, people are judicious about whom they turn to for loans, because credit relations are equally debt relationships (Taylor, 2012) and such relations are infused with social and moral dimensions that go beyond financial considerations (Guerin, 2014; Ojong & Gill-Wiehl, 2023; Peebles, 2010). Moral currency can be costly when debt and dignity collide. As a 40-year-old widow in Tanzania explained,

> I got a loan from a friend when I needed to solve that problem [purchase a SHS], and I didn't have other options. After solving my problem, I repaid my friend's money. But I do not request a loan from all my friends because some friends are like witches. What I mean is, they will tell everyone that they lent you money, and that's something I don't want to happen to me (TL, Tanzania, 2022).

This anxiety exposes how borrowing carries social risk: reputational damage can be more painful than financial loss. Credit, in this sense, is not merely an economic exchange but a negotiation over confidentiality, trust, and social standing. Households thus engage in careful relational work, selectively activating kinship ties, anticipating gossip, and distributing obligations across networks to protect dignity while accessing essential energy services (Zaloom and James, 2023). Peer lending therefore functions not only as short-term financial relief but as a moral strategy for managing precarious aspirations under conditions of constraint.

The personal cost of preserving this fragile stability is felt most acutely within kinship and friendship relations, where lending can entrench long-term hierarchies and anticipatory status conflicts. As one participant in Cameroon explained:

> In order to avoid family problems, I prefer not to request a loan from him. He [family member] might cause family problems in the future. He would say that I would not have been in a particular position if he did not assist me. The family member may not even be willing to lend money because I might end up having a better life in future, so I will be superior to him (CJ, Cameroon, 2024).

Other echoed this sentiment:

> I would feel very bad if my request for a loan was declined by a family member than if it is declined by a friend. So, I prefer to borrow money from a friend, so that if the request is declined, I know that it is just a friend (XC, Cameroon, 2024).
> I borrow money from friends. It is difficult to borrow money from a family member. As a member of a family, if I ask my brother for a loan, he will think that I will not repay the loan. But when I borrow money from someone outside the family circle, the person believes that I will repay the loan. That is why it is difficult to borrow money from a family member (LJS, Cameroon, 2024).

These intimate negotiations may appear purely local, but their effects ripple outward. A friend's loan that bridges a missed solar payment prevents a default event in a PAYG system, sustaining repayment profiles that are later bundled into international asset-backed securities. In this way, Movement I's moral economies form the scaffolding for Movement II's securitized flows. Preventing a single default in a rural community helps stabilize the reliability of repayment portfolios traded thousands of miles away, demonstrating how the most intimate acts of borrowing are already entangled with global finance.

Together, these accounts show that kinship, far from guaranteeing solidarity, can entangle borrowers in emotional hierarchies of pride, fear, and anticipated judgment.

But some participants framed kinship as a refuge rather than a hazard, revealing the ambivalence of familial finance:

> Yes, I borrow money from my cousin [...] because he is the one who has the habit of lending the amount of money I want (DZ17, Tanzania, 2022).
> We have a habit of helping one another; when I need a loan, I always call my relatives, especially my uncle who always meets my request (DW2, Tanzania, 2022).
> [...] Because my relatives know me, and I can repay the loan at any time, and they would tolerate my excuses more than others (CD17, Tanzania, 2022).

These accounts highlight kinship's forgiving elasticity. Flexibility replaces interest rates; patience replaces collateral. But even these benevolent exchanges are not free from moral accounting.

Across the excerpts, individuals calibrate decisions about whom to approach by weighing trust against exposure. Some avoid kin to escape future humiliation; others embrace kin for safety and understanding.[10] The same family tie that can nurture generosity may also generate shame. This tension exemplifies what scholars call the moral economy of debt—where obligation, reciprocity, and reputation are evaluated alongside

[10] While mainstream economics treats loans as contracts between rational maximizers (Mas-Colell et al., 1995; Varian, 2014), the evidence across Kenya, Cameroon, and Tanzania shows that credit decisions are saturated with moral reasoning.

interest and risk (Bloch & Parry, 1989; James, 2012; Ojong et al., 2025; Roitman, 2003).

The participant in Cameroon who hesitated to borrow from a cousin was performing exactly the calculus of balancing liquidity needs against the risk of relational stain. Embedding this insight clarifies why indigenous financial institutions flourish and why some households prefer a PAYG SHS lockout to public shame. The argument that economic action is deeply ethical (Sayer, 2007), therefore translates well to the solar power context: financial intermediation succeeds only when it co-opts existing moral frameworks, translating social debts into revenue streams without shredding the relational fabric that makes repayment thinkable in the first place. Friends, in contrast, may present a 'neutral zone' free from the deep emotional stakes associated with family, though that neutrality can also come with more rigid repayment expectations or uncertain trust. This suggests that peer lenders have social meanings beyond being financial intermediaries, since the credit they provide is intimately connected to future relationships, reputations, and dignity.

The everyday practices discussed here reveal that what looks like resilience or financial ingenuity is in fact the fragile scaffolding of a system designed to externalize volatility onto households, ensuring that the promise of empowerment is continually undercut by the reality of extraction and discipline.

What begins as borrowing money from a neighbor or negotiating installments with a shopkeeper is never restricted locally. These micro-acts form the very scaffolding of securitization, which I now lay bare in Movement II. The conversion of kilowatts into collateral leaves more than digital imprints. It also generates material residues such as discarded batteries, fractured solar panels, and toxic circuits that will resurface in Chapter 5. The same data infrastructures that transform energy circuits into financial assets produce ecological afterlives that connect digital value to material harm. Viewed from this lens, the securitization of light entails the pollution of soil and bodies, linking the abstraction of finance to the toxicity of its remains.

4.3 Circuits of Extraction: The Globalization of Insertion and Depletion

The practices described earlier do not remain localized; they scale upstream into global circuits structured by insertion and depletion. Micro-payments aggregated from dispersed households are inserted into securitization chains that attract international capital, while depletion is borne by the same households whose labor, liquidity, and social fabrics underwrite repayment. This globalization of insertion and depletion clarifies how Africa's solar energy frontier functions as a laboratory for financial experimentation.

While Movement I showed how households juggle obligations to keep the lights on, Movement II demonstrates that these very micro-transactions are the raw material of global finance, connecting community-level improvisations directly to securitized portfolios and distant investor logics. The four intermediary conduits revealed in Movement I (i.e., PAYG platforms, local SHS suppliers and installers, indigenous financial institutions, and peer lenders) do far more than channel solar kits from warehouses to households. Collectively, they assemble a scalar value chain that turns subsistence incomes into investable cash flows. Each mobile money top-up a Tanzanian farmer sends rides the same digital rails back to Nairobi, is aggregated with millions of other micro-payments, and is ultimately bundled into asset-backed securities marketed in major financial hubs (see Fig. 4.2 for an illustration of these macro–micro linkages). Financial capital flows from maize harvests to hedge-fund desks, while risk is displaced downstream through the ever-present threat of enforced blackout for PAYG SHS users who have yet to gain ownership of their system.

Seen through assetization and techno-economic governance (Birch, 2024; Birch & Muniesa, 2020; Roitman, 2023), PAYG solar firms have invented a novel asset class around energy-poor households (Baker, 2023). By assetization, I mean the process by which everyday goods (such as SHSs) are reconfigured into financial assets (Birch, 2024). Their value no longer derives mainly from their immediate use, but from the predictable revenue streams they can generate in the future. In the case

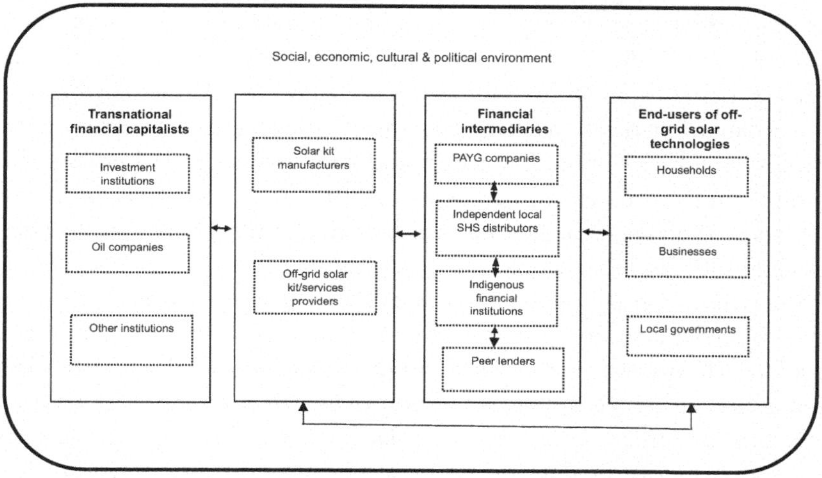

Fig. 4.2 Macro–micro circuits of financial accumulation

of SHSs, what looks like a household consumer device is simultaneously formatted as an investable financial instrument. In this context, techno-economic governance refers to the way technical devices, such as SIM-enabled lockouts, mobile-money infrastructure, and repayment dashboards, allow firms to govern households at a distance. These technologies do not just facilitate transactions; they actively shape how people use energy, when they pay, and under what conditions they are cut off. Governance here is exercised through software and infrastructure rather than face-to-face enforcement.

What began as a by-product of payment tracking is now the engine of securitization.[11] Digital client trails, first priced for broad and diverse customer data, are sliced, rated, and sold in hard currency.[12] In effect, every mobile payment a household makes toward an SHS is logged

[11] Securitization does not produce mutual benefit for creditors and debtors; rather, it is marked by asymmetrical power dynamics and exploitative financial relationships (Soederberg, 2013).

[12] These trails are not only repurposed as credit histories; they are priced as risk assets in global financial markets. As Mader et al. (2025) emphasize, default probabilities in PAYG portfolios are aggregated, securitized, and sold as de-risked investment products. In effect, the everyday struggles of households to keep up with weekly/biweekly payments become raw material for risk commodification, feeding accumulation at distant financial hubs.

and cross-referenced with phone numbers, geolocation, and repayment histories. These digital footprints are then bundled into datasets that indicate a client's reliability and purchasing power. PAYG solar firms or their financial partners package this information as credit scores or risk profiles, which can be sold to banks and data brokers in US dollars. For example, in Kenya, one woman explained that her weekly solar payments were tracked on M-Pesa, creating a history that lenders later used to assess her eligibility for other loans.

The string of small energy payments that emerges becomes a monetizable financial asset in its own right, detached from the household and circulated in international markets. In this arrangement, the sun is monetized twice: minerals upstream, digital debt downstream. The physical extraction of minerals used by the solar energy industry establishes the hardware frontier; the datafied extraction of repayment histories anchors the financial frontier. In line with Birch's (2024) argument, this transformation marks more than a shift in valuation, it signals a new regime of techno-economic governance in which future-oriented revenue streams are not just calculated but actively constituted through datafied claims. These claims, embedded in digital repayment profiles, operate as assetized expectations, recursively shaping how PAYG solar firms govern their customer base. Consequently, households are rendered classifiable not as users of solar energy, but as holders of potential revenue flows, subject to the same investor logics that define digital data within the broader asset economy.

Still, the globalization of insertion and depletion does not erase local mediations. As Movement I showed, households juggle overlapping conduits such as neighbors, indigenous financial institutions, and shop owners, whose contradictions are precisely what make securitization possible. Local intermediaries mediate, buffer, and sometimes amplify financial discipline. A single household may juggle more than one conduit in the same month, using a small loan obtained from an indigenous financial institution to buy an SHS on installment from a shopkeeper, and calling an independent installer to get the system set up. Such an overlap reveals that solar electrification-as-asset recruits existing moral economies of trust and reciprocity rather than displacing them. It is precisely because repayments are socially legible—missed installments signal shame, timely payments signal reliability—that financiers can bet on future cash flows with confidence.

But the consequences are asymmetrical. Where financial capital leaps borders, bodily and ecological costs stick to place. Fluctuating crop prices, or a child's school fee can plunge a household into darkness overnight, as the SIM-enabled controller enforces default. Far from providing autonomy, the technology renders PAYG SHS households that carry an outstanding balance vulnerable to financial shocks, since energy access is conditional on the ability to make continuous payments. As noted previously, accounts from Kenya and Tanzania indicate that families on rent-to-own SHS contracts often borrow from neighbors, indigenous savings, and credit institutions, or even go without light rather than risk arrears, showing that the mobilizing narrative of energy independence collapses under the weight of everyday indebtedness. Remote lockout is a form of micro-violence, experienced as disconnection, shame, and disruption, that secures macro-returns for distant investors. These securitized flows exemplify the 'seesaw' of global capital, in which mobile finance perpetually searches for higher returns, compelling localities to absorb risks and costs in exchange for temporary inclusion (Ortiz, 2025).

Therefore, the two Movements are braided together rather than sequential. The household improvisations documented in Movement I are inseparable from the securitized architectures of Movement II; each micro-transaction stabilizes the global circuit, and each global demand for revenue shapes the pressures households face.

In contrast to Movement I, which identified how money moves at the household and community level, Movement II reveals how household payments are siphoned outward into opaque financial circuits. Money leaves the local community and enters national treasuries, securitization vehicles, and ultimately distant financial hubs. The precise endpoints remain obscured within hedge-fund desks and investor portfolios, but what is clear is that these payments exit the household and community, becoming part of broader architectures of global finance.

Movement II also shows why the lights may be turned off for PAYG SHS households with an outstanding balance when repayment falters. Instead of simply internalizing top-down financial pressures, households actively convert digital payment streams into tools for status-building, social reciprocity, and temporal hope. Still, these practices of meaning-making are not insulated from the hierarchies and asymmetries that sustain global finance.

The bodily injuries and environmental externalities unpacked in Chapter 5, and the modalities of women's insertion under patriarchal governance analyzed in Chapter 6, converge here to show that the very stability of financial flows rests on depletion and gendered discipline. These circuits remain reliable because patriarchal governance ties repayment to women's reputational labor and care time.

4.4 Concluding Remarks

Having shown the conduits through which kilowatts become data, data become collateral, and collateral become securities, I have argued that financialization in the solar power sector is the machinery through which insertion and depletion are operationalized. What begins as household repayment ends as securitized value, while marginalized people's labor and community trust absorb the risks that investors evade. Everyday darkness, mobile-money transfers, and neighborhood lending circles together sustain the global liquidity of solar finance. In the next chapter, I extend this circuitry into its material residues, following how the solar power boom takes bodily and ecological form in Ghana's mines, solar plants, and scrapyards.

References

Anderson, S., & Baland, J. M. (2002). The economics of ROSCAs and intra-household resource allocation. *The Quarterly Journal of Economics, 117*(3), 963–995.

Ardener, S., & Burman, S. (Eds.). (1995). *Money-go-rounds: The importance of rotating savings and credit associations for women.* Berg Publishers.

Armendáriz, B., & Morduch, J. (2010). *The economics of microfinance*. MIT Press.
Baker, L. (2023). New frontiers of electricity capital: Energy access in sub-Saharan Africa. *New Political Economy, 28*(2), 206–222.
Bateman, M. (2010). *Why doesn't microfinance work?: The destructive rise of local neoliberalism.* Bloomsbury Publishing.
Bateman, M., Duvendack, M., & Loubere, N. (2019). Is fin-tech the new panacea for poverty alleviation and local development? Contesting Suri and Jack's M-Pesa findings published in Science. *Review of African Political Economy, 46*(161), 480–495.
Battaglia, M., Gulesci, S., & Madestam, A. (2019). Repayment flexibility and risk taking: Experimental evidence from credit contracts (Discussion Paper 13329). Centre for Economic Policy Research.
Beck, T., & Peria, M. S. M. (2006). *Banking services for everyone?: Barriers to bank access and use around the world* (Vol. 4079). World Bank Publications.
Bhattacharyya, S. C., & Palit, D. (2016). Mini-grid based off-grid electrification to enhance electricity access in developing countries: What policies may be required? *Energy Policy, 94*, 166–178.
Birch, K. (2024). Assetization as a mode of techno-economic governance: Knowledge, education and personal data in the UN's system of national accounts. *Economy and Society, 53*(1), 15–38.
Birch, K., & Muniesa, F. (2020). Introduction: Assetization and technoscientific capitalism. In K. Birch & F. Muniesa (Eds.), *Assetization: Turning things into assets in technoscientific capitalism* (pp. 1–40). MIT Press.
Bloch, M., & Parry, J. (1989). Introduction: Money and the morality of exchange. In J. Parry & M. Bloch (Eds.), *Money and the morality of exchange* (pp. 1–32). Cambridge University Press.
Bourdieu, P. (1984). *Distinction: A social critique of the judgment of taste.* Harvard University Press.
Bouman, F. J. A., & Hospes, O. (Eds.). (1994). *Financial landscape reconstructed: The fine art of mapping development*. Westview Press.
Brown, D., Hall, S., Martiskainen, M., & Davis, M. E. (2022). Conceptualising domestic energy service business models: A typology and policy recommendations. *Energy Policy, 161*, Article 112704.
Burlinson, A., Davillas, A., & Law, C. (2022). Pay (for it) as you go: Prepaid energy meters and the heat-or-eat dilemma. *Social Science & Medicine, 315*, Article 115498.
Cahn, P. S. (2008). Consuming class: Multilevel marketers in neoliberal Mexico. *Cultural Anthropology, 23*(3), 429–452.

Collins, D., Morduch, J., Rutherford, S., & Ruthven, O. (2009). *Portfolios of the poor: How the world's poor live on $2 a day*. Princeton University Press.

Demirgüç-Kunt, A., & Klapper, L. F. (2012). Financial inclusion in Africa: An overview. *World Bank Policy Research Working Paper* (6088).

Field, E., Pande, R., Papp, J., & Park, J. (2012). Repayment flexibility can reduce financial stress: A randomized control trial with microfinance clients in India. *PLoS ONE, 7*(9), 1–7.

Gabor, D., & Brooks, S. (2017). The digital revolution in financial inclusion: International development in the fintech era. *New Political Economy, 22*(4), 423–436.

Guerin, I., Morvant, S and Servet, J-M. (2011). "Understanding the diversity and complexity of demand for microfinance services: Lessons from informal finance." In *Handbook of microfinance*. New Jersey: World Scientific Publishing.

Guerin, I. (2014). Juggling with Debt, Social Ties, and Values: The Everyday Use of Microcredit in Rural South India. *Current Anthropology, 55*(9): 40–50.

Guyer, J. (2004). *Marginal gains: Monetary transactions in Atlantic Africa*. University of Chicago Press.

Hossein, C. S. (2024). *The banker ladies: Vanguards of solidarity economics and community-based banks*. University of Toronto Press.

Hossein, C. S., & Bonsu, S. K. (2023). Situating the West African system of collectivity: A study of Susu institutions in Ghana's urban centers. *Rethinking Marxism, 35*(1), 108–134.

James, D. (2012). Money-go-round: Personal economies of wealth, aspiration and indebtedness. *Journal of the International African Institute, 82*(1), 20–40.

Kar, S. (2013). Recovering debts: Microfinance loan officers and the work of "proxy-creditors" in India. *American Ethnologist, 40*(3), 480–493.

Kaushal, T. (2025). Gendered labour produces finance: Navigating profits and care as intermediaries of finance. *Economy and Society*, 1–24.

Khasru, S. M., & Rafee, A. A. (2021). Renewable energy empowerment at the grassroots: The success story of solar home system (SHS) in Bangladesh. https://ic-sd.org/wp-content/uploads/2020/11/Syed-Munir-Khasru.pdf

Knight, E., & Wójcik, D. (2020). FinTech, economy and space: Introduction to the special issue. *Environment and Planning a: Economy and Space, 52*(8), 1490–1497.

Kocieniewski, D., & Finch, G. (2022). How pay-as-you-go solar made the world's cheapest new energy unaffordable. *Bloomberg News*. Retrieved July 29, 2025, from https://www.bloomberg.com/news/features/2022-04-07/how-pay-as-you-go-solar-made-the-world-s-cheapest-new-energy-unaffordable

Langley, P., & Leyshon, A. (2021). The platform political economy of FinTech: Reintermediation, consolidation and capitalisation. *New Political Economy, 26*(3), 376–388.

Mader, P., Morvant-Roux, S., & Ecuyer, B. (2025). Risky waters: Household debts for financialised inclusion. *Geoforum, 165*, Article 104383.

Mas-Colell, A., Whinston, M. D., & Green, J. R. (1995). *Microeconomic theory* (Vol. 1). Oxford University Press.

Mianda, G. (2020). Francophone sub-Saharan African immigrants organizing tontines in Toronto: A basis for solidarity and integration. *Canadian Ethnic Studies, 52*(3), 7–26.

Morduch, J. (1999). The microfinance promise. *Journal of Economic Literature, 37*(4), 1569–1614.

Ockwell, D., Atela, J., Mbeva, K., Chengo, V., Byrne, R., Durrant, R., Kasprowicz, V., & Ely, A. (2019). Can pay-as-you-go, digitally enabled business models support sustainability transformations in developing countries? Outstanding questions and a theoretical basis for future research. *Sustainability, 11*(7), 2105.

Ojong, N., Gill-Wiehl, A., & Agbe, E. (2025). Cash, credit, or kin? Financing pathways and the uneven sustainability of off-grid solar electrification in Tanzania and Malawi. *Sustainable Development.* https://doi.org/10.1002/sd.70009

Ojong, N. (2018). Trust, cultural norms and financial institutions in rural communities: The case of Cameroon. *Review of Social Economy, 76*(1), 19–42.

Ojong, N. (2019a). Informal borrowing sources and uses: Insights from the North West Region Cameroon. *Third World Quarterly, 40*(9), 1730–1749.

Ojong, N. (2019b). *The everyday life of the poor in Cameroon: The role of social networks in meeting needs.* Routledge.

Ojong, N., & Gill-Wiehl, A. (2023). "A lender should not know where you live": Financial precarity, debt, and everyday life in rural Malawi and Tanzania. *Journal of Rural Studies, 97*, 314–321.

Ojong, N., & Obeng-Odoom, F. (2017). Resilience and dynamism of embedded financial transactions in Cameroon. *Journal of Economic Issues, 51*(1), 181–200.

Ojong, N., & Simba, A. (2019). Fostering micro-entrepreneurs' structural and relational social capital through microfinance. *Journal of Small Business & Entrepreneurship, 31*(1), 1–20.

Ortiz, R. J. (2025). Spatial competition and unevenness in global capitalism: Labor, nature, and the seesaw of capital. *Critical Sociology, 51*(2), 283–300.

Peebles, G. (2010). The anthropology of credit and debt. *Annual Review of Anthropology, 39*(1), 225–240.
Petry, J. (2020). From national marketplaces to global providers of financial infrastructures: Exchanges, infrastructures and structural power in global finance. *New Political Economy, 26*(4), 574–597.
Pinzur, D. (2021). Infrastructural power: Discretion and the dynamics of infrastructure in action. *Journal of Cultural Economy, 14*(6), 644–661.
Riedke, E., & Adelmann, C. (2022). The good payers: Exploring notions of ownership in the sale of pay-as-you-go solar home systems. *Energy Research & Social Science, 92*, Article 102773.
Roitman, J. (2003). Unsanctioned wealth; or, the productivity of debt in Northern Cameroon. *Public Culture, 15*(2), 211–237.
Roitman, J. (2023). Platform economies: Beyond the north–south divide. *Finance and Society, 9*(1), 1–13.
Roitman, J. (2025). Financial platforms beyond the North–South divide. In *The Cambridge global handbook of financial infrastructure* (pp. 49–63).
Roitman, J., Moon, A., & Lin, L. (2025). Digital platform economies: Value from data?. *Platforms & Society, 2.* https://doi.org/10.1177/29768624251358643
Rolffs, P., Ockwell, D., Byrne, R., Ekers, M., & Prudham, S. (2015). Beyond technology and finance: Pay-as-you-go sustainable energy access and theories of social change. *Environment and Planning A, 47*, 2609–2627.
Rutherford, S. (2000). *The poor and their money*. Oxford University Press.
Sayer, A. (2007). Moral economy as critique. *New Political Economy, 12*(2), 261–270.
Scott, J. C. (1976). *The moral economy of the peasant: Rebellion and subsistence in Southeast Asia.* Yale University Press.
Servet, J.-M. (Ed.). (1995). *Épargne et liens sociaux. Études comparées d'informalités financières [Savings and social relations: A comparative study of financial informalities]*. AEF/Aupelf.
Shipton, P. (2010). *Credit between cultures: Farmers, financiers, and misunderstanding in Africa*. Yale University Press.
Simba, A., Tajeddin, M., Dana, L. P., & Ribeiro Soriano, D. E. (2024). Deconstructing involuntary financial exclusion: A focus on African SMEs. *Small Business Economics, 62*(1), 285–305.
Soederberg, S. (2013). Universalising financial inclusion and the securitisation of development. *Third World Quarterly, 34*(4), 593–612.
Srnicek, N. (2017). *Platform capitalism.* John Wiley & Sons.

Taylor, M. (2012). The antinomies of 'financial inclusion': Debt, distress and the workings of Indian microfinance. *Journal of Agrarian Change, 12*(4), 601–610.

Tongsopit, S., Moungchareon, S., Aksornkij, A., & Potisat, T. (2016). Business models and financing options for a rapid scale-up of rooftop solar power systems in Thailand. *Energy Policy, 95*, 447–457.

Varian, H. R. (2014). *Intermediate microeconomics with calculus: A modern approach*. W. W. Norton & Company.

Weber, H. (2014). Global politics of microfinancing poverty in Asia: The case of Bangladesh unpacked. *Asian Studies Review, 38*(4), 544–563.

Yadav, P., Heynen, A. P., & Palit, D. (2019). Pay-as-you-go financing: A model for viable and widespread deployment of solar home systems in rural India. *Energy for Sustainable Development, 48*, 139–153.

Zaloom, C., & James, D. (2023). Financialization and the household. *Annual Review of Anthropology, 52*(1), 399–415.

Zhang, F. (2014). Can solar panels leapfrog power grids? The World Bank experience 1992–2009. *Renewable and Sustainable Energy Reviews, 38*, 811–820.

5
Toxic Landscapes

5.1 Introduction

At dawn in Awaso, Ghana, bauxite trucks rumble past food stalls, coating cocoyam leaves in dust. Several kilometers away, flames rise from Accra's Agbogbloshie scrapyard as discarded solar kits are stripped for copper under a haze of fumes. In between, the fences of the Kaleo-Lawra solar plant redirect women's footsteps as they walk extra miles to reach their farms or gather firewood. What seems like scattered fragments of dust, razor-wire fences, and toxic haze are in fact knots in a single circuitry of insertion and depletion that binds extraction, enclosure, and toxic afterlives. I argue that insertion is visible in the ways communities are linked to mines, solar plants, and e-waste economies as indispensable infrastructure for global green value, while depletion emerges in the erosion of lungs, soils, time, and dignity that makes this value durable. Other lenses sharpen the picture—structural violence captures the routinized injuries that slowly shorten life chances (Farmer, 2004), while necropolitics points to forms of abandonment and premature death (Mbembe, 2020). But these are not alternative explanations. They are modalities of insertion and depletion, conceptualizing how incorporation into green capitalism continually produces exhaustion.

N. Ojong, *Solar Power Capitalism*, Energy, Climate and the Environment, https://doi.org/10.1007/978-3-032-13707-4_5

Building on what I discussed in Chapter 4, how kilowatts became collateral, and how households sustained financial flows, I follow those flows into their material residues. Debt circuits there become dust plumes, fences, and toxic residues here. The three design logics outlined in Chapter 2 also resurface in embodied form: enclave privilege in the positioning of solar plants on communal land, externalization of risk in the toxic residues borne by households, and concessional underwriting in the financial architectures that legitimate such projects as development gains. Disembodied claims to 'sustainability' are materially secured through insertion into circuits of risks and depletion of already precarious lives.

The argument unfolds in three movements. The first shows how capital's entry tears landscapes apart in Awaso, Kaleo-Lawra, and Agbogbloshie. The second foregrounds the fences that enclose bodies and livelihoods once value is unlocked. The third follows abandoned solar kits and smelter dust to reveal where green externalities finally settle. Collectively, they establish that harm is not a by-product but the operating condition of investable green value.

5.2 Ghana's Green Energy Triad: Bauxite Mine, Solar Plant, Scrapyard

In the previous chapter, I discussed insertion and depletion through the discipline of household repayment. Here, they take material form in dusted lungs, fenced commons, and contaminated bodies. From bauxite mine to solar kit to scrapyard, the seemingly linear sequence of mineral-to-kilowatt-to-waste is in fact a single circuit of insertion and depletion.

At Awaso, insertion ties miners and their families into global aluminum/solar chains, even as their livelihoods rest on precarious contracts, kinship loans, and makeshift micro-enterprises. Depletion registers in the bauxite dust, the fragility of wages, and the stretching of household resilience. At Kaleo-Lawra, insertion incorporates communal land into the green energy transition by fencing it off for solar plants. Depletion appears in the everyday displacements of women's time and mobility, and in the loss of subsistence pathways. At Agbogbloshie, insertion turns e-waste workers into indispensable recyclers of solar power capitalism's toxic afterlives. Depletion registers in the burns, shortened lifespans, and poisoned soils that stabilize green accumulation.

Harvey's (2017, 1982) concepts of accumulation by dispossession and spatial fix help contextualize these dynamics. But the lived mechanics, i.e., the way bodies, ecologies, and households are incorporated and worn down, are captured most directly through insertion and depletion. These Ghanaian sites are not sequential stages but overlapping boom zones of accumulation whose volatility is the very condition of investment. Seen from this lens, extraction is not solely an input but a spatial strategy: incorporation and exhaustion fused into the very circuits that make solar energy bankable. Put differently, Ghana's green energy triad should not be considered as three separate episodes but as overlapping spheres of insertion and depletion, where the volatility of boom zones of accumulation is precisely what secures their investability.

5.2.1 Awaso Bauxite Mine

Commercial bauxite extraction at Awaso dates to the 1940s, though the deposit was first identified in 1921.[1] Control of the mine has passed through a succession of owners: British Aluminum Co. (1940–1974), a Ghanaian state-majority period (1974–1998), Rio Tinto (1998–2010), China's Bosai Minerals Group (2010–2022), and, since 2022, a Ghanaian consortium, Ofori-Poku Company Ltd.[2] Each transfer reproduced Awaso's insertion into transnational commodity circuits while diluting local residents' influence over how, and on whose terms, the mineral is removed. The mine's continuity under shifting regimes depicts how insertion endures even when ownership changes hands: what circulates steadily is not just aluminum but the depletion of soils, lungs, and livelihoods that sustain its extraction.

---

[1] https://www.ghanabauxite.com/history/

[2] https://www.ghanabauxite.com/history/

5.2.2 Kaleo-Lawra Utility-Scale Solar Plant

Ghana's flagship Kaleo-Lawra utility-scale solar plant in the Upper West Region generates roughly 35 MW of energy, enough to provide electricity to about 32,200 households (Stock et al., 2023; Volta River Authority, 2020).[3] Framed by state energy planners as a symbol of low-carbon development and climate justice, its placement also reveals a pragmatic political calculus: build where land is inexpensive, unfragmented, and easily connected to existing transmission infrastructure. In doing so, the plant materializes a familiar contradiction: renewable expansion achieved through enclosure. Communal land is reclassified as 'development' land, connecting local households to a transition that redraws access, labor, and time. Hence, insertion takes the form of infrastructural integration; depletion, the slow erosion of autonomy and subsistence.

5.2.3 Agbogbloshie E-Waste Scrapyard

Situated in Accra's urban core, Agbogbloshie has long been synonymous with the processing and disposal of e-waste, including computers, televisions, and other electronic devices imported largely from Europe and North America (Sovacool et al., 2020), as well as discarded off-grid solar (OGS) systems collected from communities across Ghana (Figs. 5.1 and 5.2). Over the past two decades, an informal recycling economy has emerged, employing young men and women who collect, dismantle, burn, and resell salvaged metals. The scrapyard is often cited as one of the most toxic e-waste sites globally (Akese, 2019; Grant et al., 2024). But the enduring label of 'world's largest e-waste dump' obscures more than it reveals: what matters are the measured toxicities and the lived harms borne daily by those who work and reside there. The dynamics at Agbogbloshie echo findings from southeastern California, where solar installations have likewise deepened inequality,

[3] The Upper West region is among the nation's most economically and energy-deprived regions (Stock et al., 2023).

Fig. 5.1 Agbogbloshie e-waste scrapyard (Author's photo taken during fieldwork)

mental-health burdens, and infrastructural neglect under the banner of green development (Dunlap et al., 2024).

In the case of Agbogbloshie e-waste scrapyard, insertion operates through indispensable but expendable labor, while depletion is written into bodies, soils, shortened futures, the bodily and ecological price of renewable energy.

5.3 Embodied Violence

Extraction may open the frontier of green capitalism, but enclosure secures it. Both are intelligible only when framed as insertion and depletion: the incorporation of communities into infrastructures of value, and the erosion of their capacities to sustain it. These harms recur across multiple registers of depletion, marking how exhaustion is lived and

Fig. 5.2 Solar panel discarded at Agbogbloshie (Author's photo taken during fieldwork)

governed. The wired fence, or the service road are not neutral infrastructures but techniques of insertion and depletion, redistributing costs onto those least able to bear them.

Solar power capitalism works by staging a double movement: opening earth to unlock value and then closing space to secure that value (see also Harvey, 2017; Stock & Birkenholtz, 2024). Whether the barrier is wired fence or service road, or a district assembly bylaw, enclosure transforms commons into conduits of capital, pushing the burdens of distance, danger, and cost onto marginalized populations least equipped to deal with them. At the same time, the fence gives rise to a precarious form of infrastructural citizenship (see Anand, 2017), in which rights to land, movement, and even electricity access depends on one's location in relation to a fortified perimeter. In what follows, I weave together two accounts: solar plant enclosures, and life in mining communities into a constellation of displacement.

At the Awaso bauxite mining site, bauxite dust pollution and poor road conditions repeatedly surface as sources of bodily harm. One woman who sells food along the roadside explained, 'the dust produced by moving cars on untarred roads worries me a lot' (F1, Ghana, 2024), echoing a concern shared by many (Fig. 5.3). Prolonged exposure to this haze has been linked directly to 'catarrh [excessive mucus build-up in the nose, throat, or sinuses] and sneezing,' alongside persistent 'eye problems and headaches' (F6, Ghana, 2024). One participant recounted it as 'the daily trauma I am facing' (F5, Ghana, 2024), revealing how constant dust and its health repercussions shape everyday experience. Every time a vehicle passes by, it contributes more fine particles to the air, leading to a buildup in overall particle concentration over time, and these minute particles can travel long distances from their original source or infiltrate homes, heightening health risks (see Barnes & Connor, 2014), settling on cloths and food, as explained by some residents:

> Because of the dust, the health of inhabitants living along the roadside is at risk, and the buildings are dirty, as dusty particles always land on the roofs and walls. The roads were initially not constructed for bauxite vehicles, but they are being used because the rails are no longer working (M2, Ghana, 2024).
>
> The mining workers produce a lot of bauxite dust during their activities, which has affected almost everything we do. For instance, the dust from the trucks plying on the road accumulates on clothes when we hang them to dry. At times, the dust enters food being prepared outside (M6, Ghana, 2024).

These accounts show how insertion into the bauxite dust economy is not limited to wage dependence; it permeates breath and food, embedding depletion into the very conditions of reproduction.

In other contexts, such as Connecticut and Massachusetts counties in the United States, exposure to PM2.5 during pregnancy has been associated with low birth weight (Bell et al., 2010). Said differently, in

Fig. 5.3 Bauxite truck on unpaved Awaso road (Author's photo taken during fieldwork)

Awaso, persistent dust pollution and airborne particulates from bauxite mining operations represent a form of structural violence that implicates even the unborn child. Babies in the womb cannot opt out of this environment; they are, quite literally, developing within it, 'displaced-in-the-womb.' Since structural violence includes the ways inequality is embodied (Dao, 2025; Farmer, 2004; Murrey & Mollett, 2023; Rodriguez et al., 2014), unborn children experience this embodiment through the maternal inhalation of bauxite dust that may compromise fetal growth. The resulting low birth weight reflects how disadvantaged conditions become physiological, placing newborns at heightened risk of infection, cognitive delays, and other health challenges. Moreover, poor maternal health services, minimal prenatal outreach, and limited economic means to relocate away from the mine area further intensify the mothers' and infants' vulnerability. This neglect is structural: it reflects a hierarchy in which corporate profits and resource extraction supersede the community's right to a healthy environment.

In addition to the health-related issues caused by bauxite dust, residents in Awaso also navigate frequent accidents stemming from 'unsafe roads,' where overloaded trucks and incessant dust reduce visibility. As one woman explained, 'The main issue is dust. Many trucks are using the roads, which poses a safety risk to pedestrians and cyclists' (F14, Ghana, 2024).

Studies conducted in other contexts contend that on dusty unpaved roads, the risk of accidents rises for drivers, cyclists, walkers, and animals due to limited visibility for both oncoming and trailing vehicles (Bhattachan et al., 2019; Jones, 2000). These impacts point to what Farmer (2004) calls structural harm in motion: poorly maintained roads and bauxite-laden vehicles endanger those with minimal power to fix, relocate, or challenge the conditions that continually jeopardize their well-being. This is the architecture of depletion: a governance of neglect that keeps extraction profitable precisely by normalizing harm.

In this mining community, people witness the relentless passage of heavy-loaded trucks on deteriorating roads, but they also experience the more insidious dimensions of structural violence: decision-making processes that never truly involve local input, a political economy that prioritizes mining revenue over residents' health, and an institutional neglect that forces individuals to cope with dusty conditions and all the attendant medical costs, most often on their own. Under such arrangements, the dust swirling around them becomes more than an environmental concern; it becomes a marker of entrenched inequities, the daily evidence of how powerlessness is codified. Many remain effectively 'displaced-in-place' (Magaramombe, 2010; Mollett, 2014; Murrey, 2015), suffering a host of preventable burdens that systematically align with deeper inequities in resource governance and public accountability.

The most telling aspect of this structural violence is that it has so many moving parts, all of which reinforce each other in ways that leave the community itself sidelined. Trucks speed through the main roads because corporations are pressed to transport raw materials, however, it is the community that absorbs the fine particle emissions, the accidents and threat of accidents, and the curtailing of everyday freedoms—mothers hesitating to let their children walk to school, roadside vendors unable to protect themselves from the perpetual haze. Paving roads or lowering truck volumes would appear to be simple interventions, but in a context like Awaso's, such interventions become politicized and stalled, revealing how deep and far-reaching the structural constraints truly are. In this setting, insertion operates through infrastructural abandonment, the selective investment that builds roads for trucks but not for people. Depletion is the everyday erosion this asymmetry produces.

In essence, the violence in motion is not just about the swirl of bauxite dust; it is also about the entrenched bureaucratic habits, the exclusionary policymaking, and the distorted priorities that keep these conditions firmly intact. For residents here, bauxite dust pollution highlights the grim reality that displacement and vulnerability need not require forced relocation; they can also be produced by the routine function of systems that treat human wellbeing as peripheral. These dynamics speak to the routine and selective neglect through which depletion operates, what Davies et al., (2017: 1263) call 'violent inaction.' In these communities, it is not simply that suffering unfolds from policy gaps or infrastructural oversight; rather, suffering becomes embedded in the very logic of solar energy expansion, precisely because no meaningful care mechanisms are designed to absorb or mitigate its fallout. 'Letting die' (Davies et al., 2017; Fassin, 2009) here is not the absence of power but its selective activation: trucks rumble past untarred roads, all while official concern remains conspicuously absent. The bodily toll is the predictable outcome of non-intervention, the kind of political indifference that sustains the solar energy frontier without assuming responsibility for its costs. In fact, necropower is not only about the exercise of lethal force, but also about governing through debilitation, sustaining life just enough to remain exploitable (Davies et al., 2017; Mbembe, 2020).

While solar energy is lauded for its minimal carbon emissions, evidence from Kaleo and Lawra uncovers unacknowledged bodily disruptions within solar plant host communities. Some residents noted how the fenced-off solar plant and newly constructed feeder roads altered daily walking routes, exposing especially women, to longer, overgrown peripheral paths.:

> The land available for cultivation has decreased because of the solar plant. Also, the time it takes to walk around the fenced solar plant before reaching the farm is exhausting. By the time I arrive, I am often too tired to work. (K-F60, Ghana, 2024)
>
> I have observed that people who rely on the river for water now must walk longer distances to fetch it since the construction of the solar plant. (K-F1, Ghana, 2024)
>
> We now find it difficult to gather firewood because most of the trees were cut down during the construction of the solar plant. As a result, we have to walk long distances to collect firewood. I sometimes have to carry it all the way from the farm, which is burdensome due to the distance. (K-F2, Ghana, 2024)
>
> People must walk longer distances to gather firewood. For example, on days when we do not go to the farm but still need firewood to cook, it takes my wife a long time to collect enough to prepare our meals. (K-M2, Ghana, 2024)

Residents hosting the solar plant experience a form of embodied violence that registers in strained bodies and reworked routines. Fenced perimeters, longer walks for water and firewood, and the loss of nearby grazing land mean that exhaustion, risk of injury, and anxiety are now woven into everyday life. These burdens reveal how solar energy infrastructure, while promising clean power, produces slow forms of displacement that redistribute vulnerability onto already precarious households.

Even as households in Kaleo and Lawra gain the promise of electricity or short-term employment, they also grapple with a reworked landscape where once-familiar routes are fenced off (Fig. 5.4). Such burdens also extend to gathering essentials like firewood. As trees have been cleared and distances between dwellings and viable firewood sources have grown, some residents shoulder the extra labor of walking long distances to collect the wood required for daily sustenance.[4] This additional workload, necessitated by the solar plant's reconfiguration of land use, is particularly depleting for those who must also manage agricultural tasks. Having to traverse bushy peripheries to gather daily necessities

[4] This point about some residents walking long distances to harvest firewood has been echoed by another study conducted in the region (Stock et al., 2023).

increases exposure to more-than-human beings, compounding anxieties about personal safety. Worries about snakebites and other dangers along these secluded paths reveal the lingering effects of 'invisible' violence, a sense of vulnerability that creeps into daily routines. Therefore, the solar energy infrastructure has created new pathways of strain, revealing how even a seemingly progressive energy initiative can produce localized forms of hardship that amount to embodied and persistent violence in the everyday lives of the host community. This is not marginal to solar energy development; it is its very architecture. The promise of clean energy depends on these invisible withdrawals of time, safety, and reproductive capacity. The slow violence of longer walks, diminished access to arable land, and heightened exposure to danger reflects a political calculus that renders certain lives governable only through abandonment. These are not inadvertent oversights but systemic omissions: forms of 'active inaction' (Tyner, 2016: 206), through which power operates by withholding care and redrawing rights to space, movement, and even access to electricity. Viewed in this light, displacement is not only spatial or economic; it is affective and political, a necropolitical sorting that consigns communities to zones of managed neglect (see also Mbembe, 2020, 1992).

This necropolitical sorting is experienced not only in abstract terms of governance but also in the everyday textures of life. I use the term 'felt violence' to denote the slow, embodied injuries that register as exhaustion, anxiety, and physiological strain even when no overt force is applied (see also Farmer, 2004; Galtung, 1969). It marks the point where structural violence becomes sensory and bodily—where displacement is lived through aching limbs, tightened breath, and reworked daily practices. The sense of embodied harm intensifies precisely at the intersection of these bodily stresses and the broader economic insecurities that shadow everyday life in places like Kaleo and Lawra. Over time, these disruptions accumulate into a 'felt violence' wherein communities endure a mounting sense of insecurity tied to diminished incomes. Beyond the challenge of lost time, longer travel or rest constraints can mean fewer opportunities to generate cash, a critical safety net for emergencies.

These embodied violences mark the human threshold of depletion. But the circuitry extends further still, entangling soil, air, and other

Fig. 5.4 Fenced Kaleo-Lawra solar plant (Author's photo taken during fieldwork)

species. The next section follows these circuits into their toxic afterlives to show how the very elements that sustain life are conscripted into solar power capitalism's endurance.

5.4 Toxic Afterlives: Insertion into E-Waste, Depletion Through Toxicity

Insertion and depletion reach their most visceral expression in solar power capitalism's afterlives. Earlier, I examined how communities were incorporated into mines and fenced landscapes, here I turn to the scrapyard and the slow poisoning of air, soil, and flesh. Insertion persists in the positioning of e-waste workers and ecologies as indispensable infrastructure for recycling broken solar systems. Depletion manifests in the degradation of bodies, rivers, and soils that absorb the residues of the so-called clean energy transition. These harms manifest across multiple registers of depletion, exposing the absence of safeguards, the calculus of abandonment, and the unequal valuation of lives that sustain renewable energy accumulation.

5.4.1 Insertion into Hazardous Recycling, Depletion as Disposable Life

The journey from bauxite particulates to e-waste fumes captures a form of violent abandonment: after aluminum is smelted and solar kilowatts converted into profit, the supply chain discards its toxic remnants in zones where regulatory oversight is weakest, and life is treated as disposable. Necropolitics re-emerges in the right to breathe, a right spatially sorted by class, income, and global trade asymmetries.

A closer look at the Agbogbloshie's e-waste scrapyard, which includes non-functional solar kits (Fig. 5.2), reveals how the country's history of dispossession, its precarious employment conditions, and its high level of poverty converge to produce tangible harm and suffering in local bodies and landscapes. In the e-waste scrapyard site, these embodied harms are frequently inseparable from daily survival strategies. One scrapyard worker explained, 'I use hand gloves and safety boots. Many women sell hand gloves around' (AB17, Ghana, 2024). This small act of self-protection embodies insertion: workers are enmeshed into circuits that demand their indispensability while offering only token defenses against depletion. While superficially a precaution, the statement simultaneously conveys the inadequacy of existing protective measures and the normalization of occupational hazards. The accounts of recycle workers that follow deepen this portrait, showing how the very gestures of self-protection become evidence of systemic neglect.

Workers 'dismantle those [scrap items] that need to be dismantled and remove the important components such as wires, sensors, and other electronic cables' (AB24, Ghana, 2024), with limited awareness or power to mitigate toxic exposures (Fig. 5.5). This exemplifies how structural violence is also a physical threat. Another community member made the link to poor environmental health explicit: 'The burning of scrap creates air pollution, which causes some community members to fall ill' (AB36, Ghana, 2024).

In this case, illness is depletion made visible: lungs and immune systems are gradually worn down as a condition for solar power capitalism's recycling flows. Pollution becomes a material embodiment of inequality because those who already exist on social and economic margins are forced into hazardous work with little choice and even less protection. These intertwined bodily and structural harms show how survival itself becomes an instrument of accumulation.

The same conditions that push individuals into salvaging solar waste, including lack of formal employment, weak regulatory oversight, and limited formal education, also confine them to enduring some of the worst occupational hazards. These toxic exposures should be framed as modalities of depletion—structural violence in the absence of safeguards,

Fig. 5.5 Worker dismantling a broken TV in Agbogbloshie (Author's photo taken during fieldwork)

necropolitics in the governance of abandonment, and racial capitalism in the marking of certain bodies as expendable.

What unfolds at Agbogbloshie is not just the recycling of e-waste but, as Murrey and Mollett (2023) argue, the operation of extractive logics: frameworks that authorize racialized and gendered violence as the very condition of accumulation. Toxic residues settle on Black bodies and ecologies, marking them as expendable in the global economy of green energy transition. Thus, insertion takes the form of coerced indispensability, while depletion inscribes exhaustion and toxicity into Black bodies and ecologies. Viewed in this light, Agbogbloshie is not an anomaly but an afterlife of solar power capitalism, where the costs of low-carbon virtue are displaced onto lungs degraded, soils poisoned, and futures foreclosed.

Here, insights from racial capitalism are crucial. Capitalism is 'never not racial' (Gilmore, 2023: 471). The disposability of Agbogbloshie's workers echoes the longer genealogy in which racial differentiation has been the ordering principle of capitalist accumulation (see Dulhunty, 2023; Issar, 2021; Melamed, 2015). Contemporary e-waste economies reproduce this logic: Black bodies become the repositories into which the refuse of global energy transition is deposited. This is racial extractivism in practice (see also Preston, 2017), where devaluation of land and life authorizes their conversion into dumping grounds for toxic technologies.

Amarillo et al. (2025) make a similar point in their study of glyphosate regimes in Colombia, where an agric-military regime produces a double glyphosate logic, one for development, another for war. What emerges is a shared global logic of toxic geographies: from aerial fumigations in Putumayo (Colombia) to solar kit incineration in Agbogbloshie, marginal bodies become sites for hosting systemic abandonment. These cannot simply be framed as environmental side effects; they are the infrastructure of depletion written in toxic materiality. Pollution here is an externalization through which green legitimacy is stabilized elsewhere.

Fumes from burning solar kits and other e-waste, together with the informal recycling of discarded electronic goods that most people in the global North never have to confront, carry pollutants linked to respiratory problems, neurological damage, and long-term health complications (Fischer et al., 2020; Li & Achal., 2020; Lin et al., 2022; Song & Li,

2014; Sovacool et al., 2020). Even babies who are not directly exposed to pollutants are not spared.[5] According to one worker, the economic benefit of the scrapyard is 'not a lot, but we are able to make a living from it' (AB37, Ghana, 2024). But making a living entails daily exposure to lead, battery acid, and other toxins. As one informal recycler noted, 'When I get the batteries from the solar home system, I use a knife to create incisions. I then empty the contents into the big drain behind me' (AB29, Ghana, 2024). With the cost of preventive measures, such as ventilation systems, protective gear, and safe disposal methods, beyond the reach of most workers and local communities, pollution becomes a signifier of who holds power and resources. Poverty channels workers into these jobs and simultaneously underwrites weak or unenforced regulations, leaving the marginalized to absorb environmental harm with no realistic avenue of escape.

But these toxins do not respect bodily borders. They cross from humans to animals, from water to crops, signaling a multispecies transmission of harm. The bauxite dust that irritates Awaso miners' bodies re-emerges as airborne toxins over Agbogbloshie, while the fenced-off solar plant in Kaleo-Lawra breeds venom that harms cattle and human alike. Violence is not an aberration at the edges of solar power capitalism; it is the remnant that ties its geographies together. To end the cycle, I must also widen the circle of victims.

5.4.2 More-than-Human Bodies: Ecological Insertion and Depletion

Toxic afterlives extend beyond human lungs and skin. Insertion incorporates soils, rivers, and more-than-human beings into hazardous recycling chains, conscripting ecologies themselves as infrastructures for absorbing waste.

As elsewhere in infrastructural and extractive frontiers (see, for example, Enns & Sneyd, 2020; Murrey, 2015), in Kaleo and Lawra, the construction of a solar plant has reshaped the local environment in ways

[5] A study of residents living near an e-waste recycling area in China found elevated concentrations of polybrominated diphenyl ethers in breast milk (Leung et al., 2010).

that not only disrupt human routines but also trigger violence toward more-than-human beings. By clearing land for the solar plant and service roads, the solar project displaced snakes, scorpions, and similar wildlife, forcing them into territories that residents considered 'community' space. One participant observed, 'Wildlife like snakes and scorpions use the solar plant as their resting place, making the area dangerous at times. We see more snakes and scorpions in the area now, compared to when the solar plant was not constructed' (K-F38, Ghana, 2024). The phrase 'making the area dangerous' signals a habitat upheaval in which wildlife finds itself exiled from original spaces and compelled to occupy human domains, thereby placing both parties into precarious coexistence. Such encounters are best understood as multispecies insertion: wildlife incorporated into human settlements as displaced infrastructures of the solar energy frontier. Multiple residents linked heightened reptile and spider visibility to habitat destruction. Their experiences show that species once active only after sundown now appear throughout the day:

> Ever since the solar plant was built, it has become common to see creatures like snakes and scorpions during the daytime. Previously, these creatures were mostly active at night (K-M5, Ghana, 2024).
> Recently, we've been seeing wildlife like scorpions and snakes around during the day. This only started after the solar plant was constructed. We used to see them before, but only at night (K-F8, Ghana, 2024).
> Snakes and scorpions have been coming into the house more frequently since the solar plant was built (K-M6, Ghana, 2024).
> Previously, it was difficult to spot snakes and scorpions during the day. But since the construction of the solar plant, it has become common to see them in daylight (K-M28, Ghana, 2024).

Residents' observations reveal how the solar plant construction has unsettled local ecologies: snakes and scorpions, once largely nocturnal, now appear in daylight and enter homes. These encounters transform everyday routines into risky negotiations with displaced wildlife, showing how renewable energy projects can generate new hazards even as they promise clean power. What emerges is a multispecies form of violence, where ecological disruption and human vulnerability are tightly entangled. As Devine and Ojeda (2017) argue, dispossession is often masked within the aesthetics of progress and infrastructure, making harm appear as a necessary collateral for development.

Interference in the natural, geological, atmospheric, and biological landscapes through the construction of the Kale-Lawra solar plant has generated new burdens for cattle in two primary ways: (1) compelling them to walk much longer distances to access suitable grazing areas, and (2) bringing them into closer contact with dangerous wildlife, specifically snakes, leading to frequent encounters and potential bites. I will elaborate on both. First, a major way in which the solar plant's construction has engendered violence toward cattle is through the sudden loss of traditional grazing spaces. As the newly fenced and secured land become off-limits, owners of cattle struggle to find open fields for their herds, forcing cattle to walk longer distances to reach pasture. One livestock owner in Kaleo emphasized how the solar plant has significantly reduced the land available for animal grazing:

> The construction of the solar plant has reduced access to grazing land. A portion of the land where animals previously grazed has been converted for the solar plant, forcing them to travel farther in search of pasture. (K-F21, Ghana, 2024).

In this setting, insertion reorganizes cattle as laboring bodies within a depleted landscape, their energy redirected to maintain pasture cycles reconfigured by enclosure.

The bigger impact is borne directly by the cattle themselves. Under preexisting conditions, short grazing journeys minimized livestock fatigue and exposure to threats; now, cattle are forced to undertake daily walks that can heighten stress and affect their overall wellbeing. An often-overlooked form of dispossession is precisely this displacement of nonhuman beings from habitual feeding grounds. Ecological harm to animals can be understood as a direct outcome of construction of the solar plant that privilege green electricity generation over multiple local ways of life, including those of cattle. Recalling that 'inequalities are embodied as differential risk' (Farmer, 2004: 305), it is evident that livestock are on the receiving end of a new structural vulnerability as they lose easy, safe access to pasture.

Second, an equally damaging dimension of this violence is that as cattle move deeper into forests, they inadvertently encroach upon the ecological spaces of other wild creatures, leading to fatalities, such as the death of cattle from snake bites, as noted by various residents:

> This is because some of the grazing lands for our animals were taken for the construction of the solar plant. As a result, the grazing land had to be relocated. Since then, people have been complaining about their livestock being stolen or bitten by snakes, leading to their deaths (K-F2, Ghana, 2024).
>
> I have lost about four cows due to snake bites. When they used to graze here (points toward the solar plant area), I never lost any. I've lost my animals because I had to relocate them to graze somewhere else (K-M7, Ghana, 2024).

These losses are depletion in ecological form: a draining of animal vitality that ripples into livelihoods, cultural reproduction, and community security. Residents link cattle losses directly to the solar plant's enclosure: animals now travel longer distances to graze, exposing them to theft and snakebites. What were once safe pastures have become hazardous frontiers, where livestock face multiplied risks. These accounts reveal how green energy infrastructure can displace more-than-human beings, producing ecological violence.

Additionally, activities in the Agbogbloshie e-waste disposal site have led to violence on nature (air, water, and soil), as described by some residents:

> The local community is heavily polluted. There is serious air and water pollution. Car tires are burned every night in the scrapyard, which contaminates the environment. Acid from batteries is poured into gutters that flow into the nearby lake. Some scrapyard workers even pour the acid directly into the lake (EWS-F3, Ghana, 2024).
>
> The community experiences air pollution from the burning of car tires at night. The water in the area is also polluted. The atmosphere alone feels toxic (EW-F5, Ghana, 2024).

> Due to the high concentration of solar PV waste, chemicals leak into water bodies, which are then consumed by both humans and animals (EG1, Ghana, 2024).
> Our community is filthy because of the scrapyard. The air we breathe and the water we bathe in are polluted (RF1, Ghana, 2024).
> The community is extremely polluted because of the scrapyard. There is air pollution, water pollution, and soil pollution (RM7, Ghana, 2024).

As noted by multiple Ghanaian residents, when e-waste is indiscriminately dumped or recycled under unsafe conditions, water bodies become targets of profound ecological harm (Fig. 5.6). At the Agbogbloshie site, acidic compounds from batteries and other harmful chemicals are repeatedly channeled into water bodies, degrading aquatic habitats and altering ecosystem dynamics (Huang et al., 2014). This violent intrusion disrupts natural processes by introducing persistent contaminants such as heavy metals, which bioaccumulate in algae, fish, and other organisms that form the foundation of aquatic food chains (Ohajinwa et al., 2018). Over time, the persistent release of pollutants lowers dissolved oxygen levels and destabilizes nutrient cycles, effectively compromising the vitality of freshwater systems (Huang et al., 2014; Kalamaras et al., 2021; Ohajinwa et al., 2018). This constitutes depletion in ecological time: a degradation that unfolds so slowly it masquerades as natural decay, disguising how green capitalism metabolizes harm into its own endurance. The quiet flow of streams turns into channels of chemical infiltration, a stark illustration of how unregulated e-waste, including solar waste disposal, enacts violence on once-thriving waterscapes. Water bodies here are inserted as repositories for hazardous residues, their depletion evident in disrupted nutrient cycles and declining aquatic life.

Soil is another crucial ecological frontier facing invasive and prolonged assaults due to informal e-waste recycling practices. The depositing and open-air dismantling of e-waste at scrap sites allow toxic substances such as lead, mercury, and cadmium to penetrate surface layers (Ohajinwa et al., 2018). Over repeated exposure, these pollutants move deeper through pore spaces in the soil, unsettling the habitat of myriad microorganisms essential to nutrient cycling and soil structure maintenance (Liu

Fig. 5.6 Polluted water body near Agbogbloshie (Author's photo taken during fieldwork)

et al., 2015). Such persistent chemical invasions degrade the soil's absorptive capacity and erode its fertility, rendering it increasingly inhospitable for native plant communities. As the contamination intensifies, the soil effectively becomes a repository for lethal contaminants, sustaining a long-term cycle of ecological harm with diminished prospects for natural regeneration.

Air, an ever-present shared resource, similarly endures acts of violence due to e-waste. As previously noted by residents, the night-time burning of car tires (Fig. 5.7) and insulated wires releases dense plumes of black smoke laden with fine particulate matter. These emissions not only disrupt the chemical balance of the lower atmosphere but also settle onto vegetation and exposed terrestrial surfaces, initiating damaging feedback cycles that can further degrade local ecosystems (see Singh et al., 2022; Stevens et al., 2020). By overloading the air with harmful byproducts, informal recycling operations transform the atmosphere into a conduit of contamination, intensifying stress on plant physiology and interrupting natural processes such as pollination and seed dispersal (see Darrall, 1989; Duque & Steffan-Dewenter, 2024; Owusu-Sekyere et al., 2024). Air, too, is incorporated as an infrastructural host for toxins—another insertion whose depletion registers in damaged ecologies and compromised reproduction of plant and insect life.

Fundamentally, human beings' intimate reliance on more-than-human beings ties their lives tightly together, such that violence inflicted on animals or ecosystems often brings subsequent harm to vulnerable communities rather than to the national political or economic elite who are complicit to the processes of violence and dispossession. When a cow is lost to snakebite in the aftermath of solar plant construction, for example, it is not the transnational financial capitalists or national economic and political elites who have vested interest in its construction who suffer the economic or socio-cultural repercussions, but the livestock

Fig. 5.7 Tires arranged for e-waste burning in Agbogbloshie (Author's photo taken during fieldwork)

owner, whose livelihood is precariously balanced on that animal. The death of cows affects a family's milk supply, income stream, and cultural traditions linked to cattle rearing. Livestock perform multiple roles: in addition to providing milk, they serve as a means of indigenous savings and credit, offer economic security and reduce vulnerability, facilitate asset accumulation, help maintain social capital, and are used to finance planned expenditures (MoFA, 2004).[6] Culturally, animals such as cattle and sheep are part of the ancestral wealth of the Dagaaba people.[7] Cattle play a vital role in marriage, which is formalized through the transfer of bride-wealth, initially in the form of cowries and later in the form of cattle.[8] Essentially, the degrading of more-than-human bodies, whether snakes displaced by cleared forests or livestock forced to forage in snake-ridden terrain, carries profound consequences for people at society's margins, those who have neither the resources to adapt quickly nor the political power to demand restitution.

Foregrounding insertion and depletion reveals that more-than-human bodies are not inert victims but active infrastructures of green energy capitalism, incorporated and worn down to stabilize renewable value elsewhere.

5.5 Concluding Remarks

In this chapter I have shown how the promise of green energy rests on landscapes of harm (bauxite dust, fences, and toxins) that anchor accumulation in the exhaustion of people and ecologies. Across Awaso, Kaleo-Lawra, and Agbogbloshie, insertion enmeshes communities into the infrastructures of extraction, while depletion registers in worn bodies, fenced commons, and poisoned soils. These are outcomes of solar power

[6] On cattle as a form of saving in other communities on the continent, see Shipton (2007).

[7] These 'sacred animals' (referred to by the Dagaabas as 'donsoglaa' or 'dongsu') belong to ancestors, and 'no person has the right to sell or slaughter a cow or a sheep for his personal use without first offering the first fruit (a bull or a ram) to the owners—his ancestors' (Kpieta & Bonye, 2012: 121).

[8] 101 Last Tribes. (n.d.). Dagaaba. Retrieved March 28, 2025, from https://www.101lasttribes.com/tribes/dagaaba.html. For a detailed discussion of the economic and cultural significance of cowries to the Dagaabas, see, Yiridoe (1995).

capitalism's spatial fix. In the next chapter, I turn from landscapes to everyday life, examining how women's (un)paid labor, financial responsibility, and bodily endurance become one of the pivotal but hidden forms of infrastructure that keeps these circuits functioning.

References

Akese, G. A. (2019). *Electronic waste (e-waste) science and advocacy at Agbogbloshie: The making and effects of "The world's largest e-waste dump"* (Doctoral dissertation, Memorial University of Newfoundland).

Amarillo, C. R., Arias-Solano, L., Ojeda, D., & Botero, S. S. (2025). One glyphosate: Toxic dispossession and the agri-military regime in Colombia. *Political Geography, 118*, Article 103288.

Anand, N. (2017). *Hydraulic city: Water and the infrastructures of citizenship in Mumbai.* Duke University Press.

Barnes, D., & Connor, B. (2014). *Managing dust on unpaved roads and airports.* Alaska University Transportation Center.

Bell, M. L., Belanger, K., Ebisu, K., Gent, J. F., Lee, H. J., Koutrakis, P., & Leaderer, B. P. (2010). Prenatal exposure to fine particulate matter and birth weight: Variations by particulate constituents and sources. *Epidemiology, 21*(6), 884–891.

Bhattachan, A., Okin, G. S., Zhang, J., Vimal, S., & Lettenmaier, D. P. (2019). Characterizing the role of wind and dust in traffic accidents in California. *GeoHealth, 3*(10), 328–336.

Dao, N. (2025). *The everyday politics of resources: Lives and landscapes in Northwest Vietnam.* Cornell University Press.

Darrall, N. M. (1989). The effect of air pollutants on physiological processes in plants. *Plant, Cell & Environment, 12*(1), 1–30.

Davies, T., Isakjee, A., & Dhesi, S. (2017). Violent inaction: The necropolitical experience of refugees in Europe. *Antipode, 49*(5), 1263–1284.

Devine, J., & Ojeda, D. (2017). Violence and dispossession in tourism development: A critical geographical approach. *Journal of Sustainable Tourism, 25*(5), 605–617.

Dulhunty, A. (2023). When extractive and racial capitalism combine–indigenous and caste based struggles with land, labour and law in India. *Geoforum, 147*, Article 103887.

Dunlap, A. A., Sovacool, B. K., & Novakovic, B. (2024). "Our town is dying:" Exploring utility-scale and rooftop solar energy injustices in Southeastern California. *Geoforum, 156*, Article 104120.

Duque, L., & Steffan-Dewenter, I. (2024). Air pollution: A threat to insect pollination. *Frontiers in Ecology and the Environment.*

Enns, C., & Sneyd, A. (2020). More-than-human infrastructural violence and infrastructural justice: A case study of the Chad-Cameroon pipeline project. *Annals of the American Association of Geographers, 111*(2), 481–497.

Farmer, P. (2004). An anthropology of structural violence. *Current Anthropology, 45*(3), 305–325.

Fassin, D. (2009). Another politics of life is possible. *Theory, Culture & Society, 26*(5), 44–60.

Fischer, D., Seidu, F., Yang, J., Felten, M. K., Garus, C., Kraus, T., Fobil, J. N., & Kaifie, A. (2020). Health consequences for e-waste workers and bystanders—a comparative cross-sectional study. *International Journal of Environmental Research and Public Health, 17*(5), 1534.

Galtung, J. (1969). Violence, peace, and peace research. *Journal of Peace Research, 6*(3), 167–191.

Gilmore, R. W. (2023). Abolition geography and the problem of innocence. *Abolition geography: Essays towards liberation* (pp. 471–495). Verso.

Grant, R., Oteng-Ababio, M., & Shin, M. (2024). Academic urban legend, Agbogbloshie: Sweeping away the "World's Largest E-Waste Dumpsite." *Habitat International, 149*, Article 103097.

Harvey, D. (2017). The 'new' imperialism: Accumulation by dispossession. In *Karl marx* (pp. 213–237). Routledge.

Harvey, D. (1982). *The limits to capital*. University of Chicago Press.

Huang, J., Nkrumah, P. N., Anim, D. O., & Mensah, E. (2014). E-waste disposal effects on the aquatic environment: Accra, Ghana. *Reviews of Environmental Contamination and Toxicology, 229*, 19–34.

Issar, S. (2021). Theorising 'racial/colonial primitive accumulation': Settler colonialism, slavery and racial capitalism. *Race & Class, 63*(1), 23–50.

Jones, D. (2000). Road dust–just a nuisance or a significant road management issue. In *Proceedings, South African Transport Conference.* Pretoria, South Africa; Washington, DC: Transportation Research Board.

Kalamaras, G., Kloukinioti, M., Antonopoulou, M., Ntaikou, I., Vlastos, D., Eleftherianos, A., & Dailianis, S. (2021). The potential risk of electronic waste disposal into aquatic media: The case of personal computer motherboards. *Toxics, 9*(7), 166.

Kpieta, A. B., & Bonye, S. Z. (2012). *Land as a "God": The gender dimensions of its wealth creation among the Dagaabas in North-Western Ghana.*

Leung, A. O., Chan, J. K., Xing, G. H., Xu, Y., Wu, S. C., Wong, C. K., et al. (2010). Body burdens of polybrominated diphenyl ethers in childbearing-aged women at an intensive electronic-waste recycling site in China. *Environmental Science and Pollution Research, 17*(7), 1300–1313.

Li, W., & Achal, V. (2020). Environmental and health impacts due to e-waste disposal in China–A review. *Science of the Total Environment, 737*, Article 139745.

Lin, S., Ali, M. U., Zheng, C., Cai, Z., & Wong, M. H. (2022). Toxic chemicals from uncontrolled e-waste recycling: Exposure, body burden, health impact. *Journal of Hazardous Materials, 426*, Article 127792.

Liu, L., Li, W., Song, W., & Guo, M. (2015). Bioaccumulation of heavy metals in soil in the e-waste dismantling area, Southeast China. *Environmental Pollution, 196*, 144–152.

Magaramombe, G. (2010). 'Displaced in place': Agrarian displacements, replacements and resettlement among farm workers in Mazowe District. *Journal of Southern African Studies, 36*(2), 361–375.

Mbembe, A. (2020). *Necropolitics.* Duke University Press.

Mbembe, A. (1992). The banality of power and the aesthetics of vulgarity in the postcolony. *Public Culture, 4*(2), 1–30.

Melamed, J. (2015). Racial capitalism. *Critical Ethnic Studies, 1*(1), 76–85.

Ministry of Food and Agriculture [MoFA]. (2004). *Livestock development in Ghana: Policies and strategies.* Animal Production Directorate, Veterinary Services Directorate, Livestock Planning & Information Unit. https://faolex.fao.org/docs/pdf/gha148233.pdf

Mollett, S. (2014). A modern paradise: Garifuna land, labor, and displacement-in-place. *Latin American Perspectives, 41*(6), 27–45.

Murrey, A., & Mollett, S. (2023). Extraction is not a metaphor: Decolonial and Black Geographies against the gendered and embodied violence of extractive logics. *Transactions of the Institute of British Geographers, 48*(4), 761–780.

Murrey, A. (2015). Narratives of life and violence along the Chad-Cameroon Oil Pipeline. *Human Geography, 8*(1), 15–39.

Ohajinwa, C. M., Van Bodegom, P. M., Vijver, M. G., & Peijnenburg, W. J. G. M. (2018). Impact of informal electronic waste recycling on metal concentrations in soils and dusts. *Environmental Research, 164*, 385–394.

Owusu-Sekyere, K., Aladago, D. A., Leverenz, D., Oteng-Ababio, M., & Kranert, M. (2024). Environmental impacts on soil and groundwater of informal e-waste recycling processes in Ghana. *Sustainability, 16*(11), 4347.

Preston, J. (2017). Racial extractivism and white settler colonialism: An examination of the Canadian tar sands mega-projects. *Cultural Studies, 31*(2–3), 353–375.

Rodriguez, A., Rodríguez, P., Saborido, M., Segovia, O., & Mires, L. (2014). Visible and invisible violence and inequality in neoliberal Santiago. *Environment and Urbanization, 26*(2), 359–372.

Shipton, P. M. (2007). *The nature of entrustment: Intimacy, exchange, and the sacred in Africa*. Yale University Press.

Singh, A. A., Eram, R., Agrawal, M., & Agrawal, S. B. (2022). Air pollution: Sources and its effects on humans and plants. *International Journal of Plant and Environment, 8*(1), 10–24.

Song, Q., & Li, J. (2014). A systematic review of the human body burden of e-waste exposure in China. *Environment International, 68*, 82–93.

Sovacool, B. K., Hook, A., Martiskainen, M., Brock, A., & Turnheim, B. (2020). The decarbonisation divide: Contextualizing landscapes of low-carbon exploitation and toxicity in Africa. *Global Environmental Change, 60*, Article 102028.

Stevens, C. J., Bell, J. N. B., Brimblecombe, P., Clark, C. M., Dise, N. B., Fowler, D., Lovett, G. M., & Wolseley, P. A. (2020). The impact of air pollution on terrestrial managed and natural vegetation. *Philosophical Transactions of the Royal Society A, 378*(2183), 20190317.

Stock, R., & Birkenholtz, T. (2024). Tracking the sun: Exposing India's solar dispossessions. *Globalizations*, 1–20.

Stock, R., Nyantakyi-Frimpong, H., Antwi-Agyei, P., & Yeleliere, E. (2023). Volta photovoltaics: Ruptures in resource access as gendered injustices for solar energy in Ghana. *Energy Research & Social Science, 103*, Article 103222.

Tyner, J. A. (2016). *Violence in capitalism: Devaluing life in an age of responsibility*. University of Nebraska Press.

Volta River Authority. (2020). 35MW Solar Power Project: Upper West Regional Project Sites – Non-Technical Summary of EIA Report. Retrieved on March 28, 2025 from https://www.vra.com/media/2020/pdd/35MW%20Solar%20Power%20Project%20-%20Upper%20West%20Regional%20Sites%20%20Non%20Technical%20Summary%20Report.pdf

Yiridoe, E. (1995). Economic and sociocultural aspects of cowrie currency of the Dagaaba of Northwestern Ghana. *Nordic Journal of African Studies, 4*(2), 16–16.

6

Gendered Infrastructures

6.1 Introduction

The same practices that deliver light also reproduce gendered hierarchies of obligation, authority, and risk. I argue that solar power technologies (re)organize who pays, who decides, and who bears the everyday stresses of keeping lights on. Drawing on cases across Ghana, Kenya, and Tanzania, I show how women are repeatedly positioned to make the system work: they initiate purchases, manage installments, stretch food budgets when payments are due, walk farther when land is fenced, and accept hazardous piece-rate tasks in scrap economies. These processes operate through patriarchal orders at household, community, and workplace scales.

Following Hunnicutt (2009), patriarchy here refers to differentiated, context-specific systems embedded in distinct historical and institutional arrangements through which obligations, risks, and entitlements are allocated.[1] Within this shifting architecture, I use insertion to denote the

[1] For a rich discussion of patriarchy in the African context, see Amone et al., 2021; Bawa, 2012; Dogo, 2014; Sanya, 2013.

N. Ojong, *Solar Power Capitalism*, Energy, Climate and the Environment,
https://doi.org/10.1007/978-3-032-13707-4_6

conversion of women's financial ingenuity, embodied care, and household credit into the infrastructure that stabilizes solar energy markets, a conversion administered through domestic hierarchies, land deals, and precarious work. From this perspective, insertion contributes to broader debates on capitalism as a system animated through difference. As Omstedt and Ebner (2024) argue, uneven development cannot be disentangled from the production of social difference; capitalism is best grasped as a matrix in which spatial and gendered hierarchies are actively constituted and mobilized. Extending this insight to energy transitions, I show how women's labor, time, and credit are transformed into infrastructural stabilizers of solar energy markets.

Patriarchal governance renders women's everyday work such as stretching budgets, rerouting care routines, or handling hazardous scraps foundational to solar power capitalism's stability. Insertion marks the conversion of that labor into solar value; depletion captures the slow erosion of time, dignity, and bodily health that sustains accumulation. It is in this sense that insertion provides the analytical anchor for the chapter.

The discussion proceeds through three interlocking spheres of women's incorporation. First, household finance, where pay-as-you-go (PAYG) solar home system (SHS) contracts conscript women as financial managers and moral guarantors. Second, communal enclosure, where utility-scale solar plants impose what I call a 'gendered time tax,' forcing longer walks for firewood, water, and farmland. Third, toxic work in solar waste economies, where women's piece-rate labor renders discarded devices profitable under hazardous conditions. Collectively, these spheres reveal how solar energy capitalism reorganizes women's financial responsibilities, spatial mobility, and bodily risks so that cash flow, land, and scrap can be turned into reliable returns.

Across these sites, women also describe tangible benefits, including cleaner air, children's study time, and pride in electrified homes; thus, each sphere must be read not only as a site of burden but also of ambivalent gain. As one Kenyan female resident reflected, 'I can only thank them [the SHS company] because they brought me out of darkness, and now I have light' (LZH, Kenya, 2024). Women highlight relief from kerosene smoke, the ability for children to study at night, and, in places like Lawra, the opportunity to sell cooked food and cold drinks in the evenings. A Lawra participant noted, 'Some women are now into the sale of frozen fish and meat [...] others sell food items in the night. These could not have been possible previously since there was darkness' (UDJ, Ghana, 2024). Similarly, a Tanzanian mother of four observed, 'We get light, watch the news, and use a blender' (PIV, Tanzania, 2022). Such improvements touch on health, dignity, and educational possibility, but the very benefits women prize are secured through depletion, as the stretching of budgets, time, and bodies becomes the hidden subsidy of solar energy returns. What is framed as empowerment is therefore inseparable from the slow exhaustion that keeps the system functional.

The analysis that follows treats this contradiction as central. Solar power improves life in ways women value even as it enmeshes them more tightly in debt, enclosure, and unpaid labor (see also Fraser, 2016). Across the three spheres, gendered rhythms of time and debt are converted into stabilizers of solar returns under patriarchal governance. That contradiction is a defining feature of insertion. Indeed, the valorization of women as inherently responsible financial actors reflects a global inclusion agenda in which gender equality is pursued less as social justice than as a strategy for extending financial reach (Kaushal & Predmore, 2025). These insertions are the everyday enactments of the design logics outlined in Chapter 2: enclave privilege that diverts women's labor around fenced solar plants; externalization of risk that ties their credit and reputations to household liquidity; and concessional underwriting that renders unpaid care and labor the hidden subsidy of solar energy markets.

6.2 Governing Energy Through Gender

Conventional literature often casts off-grid solar (OGS) electrification as contributing to women's empowerment (Khandker et al., 2014; Lemaire, 2018). In reality, the PAYG SHS model uses women as the everyday gatekeepers of energy access, responsible for keeping installments flowing, negotiating small loans, and absorbing blame if the light goes off. Such dynamics parallel Kaushal's (2025) observation that women microfinance staff perform a dual role: generating financial returns while sustaining the social relations that make repayment legible, a reminder that the disciplining force of solar finance is inseparable from the feminized labor that animates it.

Benya's (2015) work on 'invisible hands' in Marikana helps clarify that these burdens cannot be framed as simply household matters but key inputs into accumulation regimes that externalize the costs of household survival onto women. In PAYG SHS contracts, the system and its lockout function both enable installments and localize accountability to women as proximate overseers of energy continuity. These conditions exemplify what Bouzarovski and Petrova (2015) describe as 'energy precarity,' a systemic condition produced by spatial inequalities, infrastructural neglect, and privatized energy delivery models.

In East Africa, mobile phones and mobile banking have become arenas where everyday gender norms are contested and reworked, even as they fold women into new infrastructures of monitoring and payment (Sanya, 2013). Energy innovation is structured precisely in this way. Empowerment through access is, in fact, insertion through surveillance and depletion through domestic austerity. For women, this ambivalence is manifested in daily micro-decisions about which bills to pay, which meals to skip, and which routines to forego. As one Kenyan woman put it, 'I was finding it difficult, contemplating whether to buy soap, or pay for solar' (QDX, Kenya, 2024). The PAYG SHS disciplines the household by reframing agency as debt maintenance. This is the bargain on offer: women accept intensified monitoring and reputational risk in exchange for light, status, and children's study time.

Women were often the initiators of SHS adoption. I conceptualize this as household financial governance rather than as care labor. A Malawian female participant explained, 'I decided to buy a solar home system to reduce the cost of batteries for torches, which had to be replaced frequently' (PYE, Malawi, 2022). The work here is budgeting, bargaining, and reputational management within unequal domestic hierarchies.

In Tanzania and Kenya, women approached local vendors, used indigenous financial institutions to pool deposits, or borrowed from friends to meet down payments. They framed SHS purchases strategically, whether by emphasizing economic benefits for the household or by invoking children's education, to recruit male consent. As one 24-year-old casual laborer in Kenya noted, 'Buying kerosene every day or using candles was a burden… I felt that if I bought the SHS it would help me in phone charging and lighting' (DII04, Kenya, 2023). Others pitched it as a status upgrade, one that could reposition the household within local hierarchies of modernity and aspiration. One Tanzanian woman framed the decision in practical and aspirational terms: 'I was the one who decided to buy the SHS because I needed light. Sometimes kerosene was too expensive or unavailable. I also wanted to try new technology' (CD08, Tanzania, 2022).

A parallel tension surfaces in rural Kenya: 'I am always in the kitchen, and candles burn quickly and finish. Sometimes, candles cost KES30 [US$0.23] per day. So, I told my husband that instead of buying candles, we should get a SHS' (YII03, Kenya, 2023). Women must manage the repayment schedule, ensure sufficient liquidity in mobile accounts, juggle food costs against installments, and deal with reputational damage if the light cuts off due to late payment, or when lenders publicly demand repayment. Sometimes, someone [lender] is shouting 'I want my money back' in front of friends and children' (RY05, Tanzania, 2022). These negotiations reveal that household decisions about solar energy are both economic and moral acts situated within patriarchal hierarchies.

These decisions exemplify bargaining with patriarchy (Kandiyoti, 1988; Kawarazuka et al., 2019), where women stretch household hierarchies to secure light, status, and study time for children while accepting intensified responsibility for repayment. These bargains with patriarchy are economic technologies of insertion—disciplinary arrangements that convert care into collateral.

While repayments weigh heavily on women, many also emphasize that the shift away from kerosene frees household income and improves health, a point underscored by one Kenyan participant, 'Previously, when we used kerosene lamps, the smoke from the lamps would cause health problems for the children. So, solar power is good because it produces no smoke, it's free from causing diseases' (EQJ, Kenya, 2024). The attraction of solar energy lies precisely in its double character: as both relief and burden.

This ongoing management of solar power access is neither invisible nor passive. It demands everyday acts of calculation, sacrifice, and improvisation. One Tanzanian woman explained how she divided her weekly market income into three envelopes: one for food, one for the SHS, and one for school fees. When sales were poor, the SHS envelope came first, 'because if the light goes, the shame is too much' (TF1, Tanzania, 2022). This account is not isolated. It points to a widespread reality in which women's labor and money management skills are repeatedly mobilized to sustain PAYG solar energy access.

Such choices often reflect a genuine weighing of gains. As noted earlier, several participants described pride in providing evening light and satisfaction that children could do homework, sometimes framed as the greatest reward of all. 'Because I have light throughout the night, my kids are able to study well, which makes me happier than anything else' (JQN, Tanzania, 2022). These are meaningful improvements, analytically distinct from, yet entangled with, debt. These intertwined emotions and obligations expose the material and psychological costs that underwrite solar energy markets.

These acts of improvisation and care are part of a broader pattern in which women absorb the costs of keeping the lights on. The strain goes beyond budgeting. The depletion (as theorized in Chapter 1) is material, emotional, and symbolic. Women absorb the stress of potential lockouts if they default on their rent-to-own SHS payment, manage the shame of returning to kerosene, and regulate household usage to avoid overloading the battery. In many cases, they impose limits on lighting or restrict charging privileges for mobile phones, decisions that confer to them authority but also make them the face of energy discipline within the home. In the case of PAYG solar energy adoption, women's budgeting, coordination, and domestic discipline are foundational to the functioning of that market, but they are seldom recognized as such. Instead, these labors are naturalized as maternal duty or financial prudence.

These domestic negotiations are mechanisms through which insertion materializes at the household scale, transforming women's budgeting and care into the infrastructure of repayment. Each sacrifice, delay, and adjustment extend depletion as the silent cost of solar power stability.

Still, if insertion operates through debt and discipline, it also unfolds in a sphere of negotiation. Women are not only bearers of depletion but also tactical actors who maneuver within, and sometimes against, the contractual architecture.

It is important, however, not to collapse these processes into victimhood. Women's engagement with SHSs is also marked by forms of situated agency. Some selectively delayed payments to free up capital for urgent needs, knowing that the first lockout was only temporary. Others deliberately restricted children's use of lighting to conserve power, asserting control over how solar power was consumed. This form of domestic energy governance reflects women's embedded energy expertise, knowledge born from lived necessity, accumulated through everyday negotiations with chronic scarcity, care obligations, and resource management (Standal & Winther, 2016). This is structural maintenance: women's adaptive intelligence maintains a financial machine that depends on their exhaustion.

The lockout function of PAYG SHSs converts darkness into an enforcement tool, the emblematic move of insertion. Women often preempted disconnection by repurposing savings, borrowing from neighbors, or delaying other purchases. Agency was present, but it was confined by a contractual architecture that externalized risk onto households.

The PAYG SHS-household becomes a kind of micro-utility, run by women, financed through labor, and governed by mobile-money infrastructures. This arrangement keeps the lights on but also keeps financial capital flowing: each repayment feeds investor portfolios, each default signal updates credit profile. The value of solar energy is not just in the electricity delivered, but in the routines and responsibilities it anchors inside the home. OGS energy access is, in fact, a reconfiguration of domestic labor into a risk management strategy for global finance. Hence, the SHS is a bridge between global capital and gendered care. Women's labor bridges that divide, an infrastructural achievement rendered possible not through technology alone, but through the quiet, recursive work of managing uncertainty.

It is true that these everyday calculations are financial survival strategies, but they slowly drain bodily energy and emotional reserves, setting the stage for the wider depletion discussed in the next sections.

6.3 Enclosure, Time Taxes, and Patriarchal Spatial Orders

In Ghana, the Kaleo-Lawra utility-scale solar plant created spatial orders that redistribute time and effort by gender. The same calculus of externalized risk re-emerges, magnified, when we move from solar goods to communal lands. While SHSs rewire financial obligation inside the home, the utility-scale solar plant reorders land and labor at community scale through male-dominated decision channels that routinize the de-prioritization of women's use-based claims. These shifts are stabilized by patriarchal governance, since chiefs, elders, and household heads recognized as legitimate decision-makers are predominantly men (Dankwa, 2018), while women's use-based claims are routed through them and de-prioritized in practice, with compensation channels routed through chiefs and elders while women's user rights remain informal and unpriced.

This is not a residue of patriarchy but an operational logic: the governance of land doubles as the governance of women's time. This is the material continuation of Chapter 5's enclosure and displacement dynamic: I now focus on how those fences reshape care work and time poverty for women.[2] In communities surrounding the Kaleo-Lawra solar infrastructures, women describe the toll of enclosure in bodily terms: longer walks, heavier loads, unfamiliar paths. The shift is not just spatial but temporal and affective. A 55-year-old mother of six put it more starkly: 'We now find it difficult to gather firewood because most of the trees were cut down during the construction of the solar plant. [...] I'm

[2] See also Klingler et al., 2024; Stock & Birkenholtz, 2024; Yenneti et al., 2016 on solar-land convergence.

growing older, and carrying such heavy loads takes a toll on my body, especially my back and waist' (K-U31, Ghana, 2024).

This is a quintessential 'time tax,' an unpriced, recurring drain on women's labor that never appears on the project's reckoning costs. What solar energy investors log as clean megawatts translates, for many women, into depleted time, strained care routines, and curtailed autonomy over land and livelihood. These dispossessions are seldom framed as such. Because land acquisition is negotiated via traditional leaders and paid in lump sums, the downstream effects on gendered labor disappear from cost–benefit calculations. However, it is precisely in the gaps between land loss and labor rearrangement that solar energy capitalism reproduces its quietest harm. Crucially, the time tax is produced by male-dominated land allocation and compensation channels that price land once and women's labor never.

Communal land, once used for firewood collection, is now fenced off, its borders guarded not just by wires but by new logics of exclusion. This affects women disproportionately, as they are the ones tasked with procuring firewood, fetching water, and ensuring daily subsistence. One woman in the Kaleo area described walking additional miles to collect firewood from more distant land. These experiences suggest a form of displacement that is both partial and permanent, what scholars have framed as being 'displaced-at-home' (Magaramombe, 2010; Mollett, 2014; Murrey, 2015). These displacements are not anomalies of implementation; they are the everyday expressions of insertion, where women's mobility becomes the silent subsidy of green energy accumulation.

This framing helps reveal why compensation packages that price land once but never account for recurring time taxes systematically misrecognize women's losses. But women also recognize infrastructural improvements that accompanied these projects. In Kaleo, residents pointed to new feeder roads and a renovated school as tangible community benefits, useful works that nonetheless misprice women's time. These infrastructural benefits do not negate women's losses; rather, they exemplify the contradictory calculus of insertion, where improvements in roads and schools are secured precisely by offloading new time taxes onto women's bodies. This resonates with Ossome's (2024) argument that agrarian dispossession is not merely an economic process but a gendered reordering of labor, whereby women's access to commons and reproductive resources is silently eroded under the guise of national development. This is depletion rendered invisible by progress—women's unpaid adjustments are the hidden costs that make solar power infrastructure legible as success.

These shifts mark another register of insertion: women's time and mobility are redirected to uphold accumulation, while depletion unfolds through the unrecorded erosion of bodily energy, dignity, and ecological knowledge.

Depletion here operates on multiple registers: ecological, bodily, temporal, and it settles unevenly on those least positioned to refuse.[3] While Rai et al., (2014) conceptualize depletion as an imbalance between the inputs and outputs of care labor, Lingham and Johnston (2025) extend this to show that depletion is not simply a consequence, but a structural logic, produced through systemic violence, infrastructural abandonment, and institutional neglect. This framing reveals how development regimes treat women's labor as infinitely elastic, using exhaustion as a governance strategy rather than a failure of design.

These dynamics materialize in the Kaleo-Lawra solar plant development, where spatial fixes, what Chapter 1 describes as the drive to secure cheap, flexible land under minimal governance, are enacted through land negotiations that exclude those most vulnerable to depletion. Local chiefs were incentivized to release communal land through compensation and vague promises of community benefits. Women, however, were rarely consulted. Their user rights were assumed to fall under patriarchal authority, rendering them invisible when land access was fundamentally restructured. Depletion here is not only material, involving longer walks, lost gathering spaces, and intensified labor, but also political, as women's claims to land and livelihood are eroded without ever being formally denied.

It is telling that while most men interviewed spoke of solar power as 'progress,' women's accounts returned repeatedly to the burdens it introduced, silent, unpaid, and structurally embedded in the political

---

[3] For a detailed discussion of an alternative conceptualization of 'depletion,' see Rai et al., (2014). For a link between the exploitation of life and the exploitation of nature, see Barca (2020).

economy of solar electrification. In addition to symbolizing development, the solar plant is reordering women's daily geographies in ways that are legible as progress to energy planners but experienced as lost time by those who must reroute.

This gendered asymmetry is reinforced not just through land and labor, but through infrastructure's aesthetic. Solar plants appeared orderly and futuristic, grids of gleaming panels stretched across neatly cleared land. Their visual promise of modernity conceals the social rearrangements they necessitate.

As one man in Kaleo put it, 'The project has made the village a tourist site. It attracts people from different places who come to visit the solar plant. This allows us to meet new people from different communities, and it also beautifies the place at night' (JWY, Ghana, 2024). Another man noted, 'The community is looking bright and busy. All trading activities are going well. The community is also a tourist site to other nearby communities. It makes us feel important and special' (YTX, Ghana, 2024). These accounts indicate how the solar plant's infrastructural spectacle is translated locally into narratives of visibility, pride, and belonging.

Yet this rebranding of space substitutes appearance for equity: some men narrate infrastructural spectacle and status; some women narrate longer routes, heavier loads, and eroded commons. As one man observed, 'Many people and groups travel from as far as Accra to view the project. The dressing of the women visitors to the community has been influential on our women' (VZV, Ghana, 2024). The gendered framings of beauty and progress reveal a deeper ideological work at play.

In this formulation, the solar plant does not signal destruction; it symbolizes progress. Still, beneath the surface, the solar plant erases access, redistributes exhaustion, and remakes rural geographies in the image of speculative finance. Kaleo and Lawra become valuable not for what they produce, but for what it enables: returns on solar energy investment, packaged for consumption in major financial hubs around the world.

Beyond labor time, enclosure also erodes gendered ecologies of knowledge. As one 33-year-old woman explained, 'Before the solar plant was constructed, there were shea trees in this area. […] We now have to go much farther into the bush to find shea nuts and gather the roots' (K-F10, Ghana, 2024). These gathering practices are more than subsistence routines; they are conduits of intergenerational knowledge transfer and repositories of cultural and medicinal expertise (see Ochieng' Obado et al., 1995 Odore, 2002; Ouma, 2022). In this context, the shea tree represents more than natural resource; it is part of a gendered landscape where women act as stewards of both ecology and tradition. Solar power development, by clearing such spaces, quietly dismembers these roles.

Nevertheless, the narrative of progress remains intact. One man noted a benefit: 'As a result of the power provided by the solar plant, the area remains well-lit at night, enabling women to carry out other activities' (K-F13, Ghana, 2024). But access to solar electrification does not replace what was lost, it redirects women's labor into other necessary but unremunerated tasks. Some women did access wage work during construction: 'Some women in the community were employed to cook for the workers, and others were employed as cleaners' (K-F19, Ghana, 2024). Still, even these opportunities were filtered through unequal channels.

As one resident recounted, 'The political parties and the assemblyman who are in charge of employment for the solar plant premise sent only their relatives for the positions' (L-W2, Ghana, 2024). What emerges, then, is not an account of empowerment, but a redistribution of light, labor, and loss.[4] In this context, patriarchal bargain trades infrastructural spectacle and feeder roads for longer routes, eroded commons, and compressed care schedules. These gendered displacements prefigure the wider argument of the next chapter: that the very insertions enabling solar power accumulation, ultimately generate the bodily exhaustion and ecological strain that constitute depletion's core logic.

At the end of the chain, some solar systems that once disciplined repayment reappear as waste that disciplines bodies, a final, lethal form of insertion. The cumulative effect of longer walks, eroded commons, and compressed care routines is lived depletion, foreshadowing the toxic afterlives I lay bare in the next section.

---

[4] On women and the complexity of empowerment in an African setting, see Bawa (2016).

6.4 Toxic Work

In the e-waste scrap economy, women are wage seekers under piece-rate logics; this is productive, not reproductive, labor. Conflating it with social reproduction obscures how workplace governance extracts value through hazard and output targets. Insertion takes the form of hazardous wage-seeking that makes solar power's end-of-life materials profitable under patriarchal wage-labor governance.

In Chapter 5, I ended with a warning about toxic afterlives. Here, I pick up the trail, as I follow e-waste into Agbogbloshie and into women's bodies. The solar systems, once decommissioned, do not vanish. In Ghana, they morph into toxic remnants such as solar panels, broken wiring, and degraded batteries, often rerouted into the Agbogbloshie e-waste scrapyard, despite Ghana's Hazardous and Electronic Waste Control Act 917 (2016), whose enforcement remains piecemeal. At the tail end of solar systems' supply chain lies a hidden circuit of death-bound value, where waste is not discarded but repurposed into precarious livelihoods. Thus, I follow the solar kits' afterlife and reveal how women's bodies bear the costs of necro-accumulation: value extraction from life-depleting processes.

Discarded solar systems arrive alongside other e-waste: once hailed as liberatory technologies, they now resurface as toxic waste. Covers are torn apart for copper wires and other materials. Informal recyclers, including women and children, sort, smash, and smelt components with rudimentary tools, inhaling fumes and absorbing long-term toxicants. The Agbogbloshie economy thrives on this necro-value: wealth generated through bodily depletion, precarious survival, and ecological decay (see also Akese & Little, 2018). In this area, work and e-waste are inseparable.

In the household economy section, I examined depletion through care; in the scrapyard economy, I analyze exploitation through wage, piece rates, and hazard. Women dig through heaps of electronics under a haze of smoke, scavenging components that can be sold by weight. Some work for middlemen who set daily quotas; others operate independently, bartering for scraps they hope will fetch a few cedis. One female recycler in Agbogbloshie explained, 'We don't know what is inside [the devices], but we know copper is life' (TRX, Ghana, 2024). That kind of knowledge, felt through the hands, built over time, and tested in daily life, is survival know-how.

But it is also evidence of structural abandonment. No formal health protection exists. State-owned medical facilities are poorly equipped and overwhelmed. Respiratory illnesses, skin ailments, and reproductive problems proliferate. For some women, the scrapyard also represents opportunity: a place to earn cash income, gain respect in their families, and sustain household needs. These openings do not cancel the risks of acid burns or toxic smoke, but they reveal the contradictory ways solar home systems afterlives are woven into women's economic survival strategies.

Toxic afterlives intersect gender. Women at the Agbogbloshie e-waste site repeatedly emphasize the overlapping burdens of childcare and income generation. A 31-year-old mother with no formal education stated, 'I have five children, and working here helps me cater to them' (EW_F1, Ghana, 2024), highlighting how the demands of care and economic necessity are intertwined. The nexus of childcare and income generation illuminates how gendered expectations compound the vulnerability of women, effectively forcing them to reconcile familial obligations with the demands of an informal labor market.

These dynamics resonate with decolonial feminist analyses of capitalism, which reveal how care and reproductive labor is systematically racialized, feminized, and instrumentalized to sustain global economic life (Bhandar, 2018; Bhattacharyya, 2018; Ossome, 2024; Vergès, 2021). In such contexts, the Agbogbloshie e-waste site emerges as both a site of economic opportunity and an arena of socio-cultural contestation, where the need to provide for one's family is inextricably linked to broader patterns of gendered survival and marginalization.

However, the provision of care is not without its hazards. Environmental toxicity exacerbates these challenges. Some women lament the prevalence of toxic fumes and frequent accidents, which threaten both their health and family stability. Their depletion is tied to exploitation as informal workers at the bottom of a class hierarchy. On this point, Amma (all names here are pseudonyms), a 43-year-old mother of two confided,

> I came to Accra to hustle, but when I got here, the only job I could find was prostitution, which went against my will. So, I decided to start picking scraps to sell. Over time, that led me to secure a space in the scrapyard, where I began buying and selling scrap materials. I was admitted for breathing difficulty in 2021 [...] the doctor thought I was a smoker, but it was because I inhale smoke every day at the scrapyard (EW, Ghana, 2024).

This excerpt exemplifies how insertion mutates at the e-waste frontier: what once disciplined households through repayment now disciplines bodies through toxic exposure and precarious wage labor. Her experience reveals a troubling intersection between environmental risks and gendered expectations of household provision, effectively placing women in a double bind, pressured to secure an income while navigating acute health dangers that could destabilize family life. The inhalation of toxic fumes and particulate matter at scrapyards like Agbogbloshie is not just a personal health risk but a systemic threat to the very foundations of family survival and intergenerational wellbeing. When women, often central caregivers, fall ill, the burden of childcare, household maintenance, and emotional labor either collapses or is transferred to older girl children or other overburdened women, fracturing already fragile support systems (Federici, 2020; Hochschild & Machung, 2012).

Here, insertion acquires its most corporeal form: the body itself becomes the conduit of value, while depletion is inscribed in lungs and skin.

Capitalist accumulation often externalizes the costs of care onto households, disproportionately burdening women (Fernandes et al., 2023). In this way, environmental precarity becomes a direct axis of reproductive precarity, as the depletion of women's bodies through unsafe labor undermines their capacity to sustain life at home. As Lingham and Johnston (2025) argue, depletion should be understood not as the sum of unpaid labor, but as the cumulative impact of structural violence, including environmental, economic, and gender-based violence, which renders women's capacity to sustain households and livelihoods increasingly unsustainable. What emerges in Agbogbloshie is a concentrated site of such violence, where sustainability is materially disarticulated from social wellbeing.

Moreover, in contexts where public health infrastructure is weak or inaccessible, illness often goes untreated, leading to cycles of incapacitation and loss of livelihood. These health risks actively threaten the daily and generational capacity of women to sustain work and family life. They deepen both economic and social insecurity in ways that remain largely invisible in conventional labor narratives. As recent feminist analyses of depletion in conflict zones reveal (Lingham & Johnston, 2025), the erosion of reproductive capacity is systemic—whether through war, infrastructure withdrawal, or environmental degradation. Solar power capitalism, while distinct, follows similar logics: it displaces risk and extracts value by exhausting the very labor that sustains life. Recognizing this dynamic forces us to ask not only who benefits from solar electrification, but who is being quietly depleted in the process. The following narrative of a female informal e-waste recycler typifies this necropolitical risk transfer, where bodily depletion underwrites circular-economy rhetoric.

> I dismantle metal scraps for my boss. I only get paid for dismantling. My boss sells the parts to bigger scrap companies. I earn GHS1,000 (US$82) monthly. I handle solar devices almost every three days. I separate the lead from solar batteries. My boss sells the lead to a repairer, while the other components are sold to an iron rod company […] I use a metal hammer and a machete to cut open the solar batteries and remove the lead. Aside from wearing gloves and safety boots, I have no other protection against toxic substances. I just have to be extra careful not to hurt myself, because the bosses do not care if I get injured […] It's emotionally draining; they speak to you disrespectfully. I always feel exhausted because of the heavy workload. I'm even planning to stop if I find a better job […] I once stepped on a sharp piece of metal that cut through my boot and severed my right big toe. I had to survive on hospital food, which was barely enough. Some good Samaritans visited and brought me food and even paid for my hospital bills. As for my family, I couldn't support them at all during that time (EF5, Ghana, 2024).[5]

[5] The exchange rate used throughout this book is US$1 = GHS12.

The hazard intensity is high, the bargaining power is low, and the wage is contingent. Women's insertion here is as quasi proletarians whose bodies absorb costs that keep the value of scrap metal realizable. The bargain trades short-term relief for enduring harm.

The extract from the female scrap worker is a window into the gendered political economy of green value extraction. Her labor is indispensable and invisibilized at once. The dismantling of solar batteries was not simply hazardous labor; it was market-sustaining work that reproduced the green capitalist economy itself. By extracting valuable components from discarded devices, she enables the circulation and reuse of materials deemed essential to the low-carbon energy transition. Nevertheless, her labor remains low paid, precarious, and dispossessed of recognition. Capitalist production relies on feminized maintenance and reproduction work, particularly the labor of women (Bhattacharya, 2017; Federici, 2012). In this case, the green economy does not transcend the exploitative logics of traditional capitalism; it retools them, mobilizing women's labor not as empowerment but as cheap, expendable, and risky.

Like the work of several other women at the e-waste site, her informal, extractive, and hazardous labor exemplifies the gendered circuits of value that underpin solar power capitalism in the global South. Her description of using rudimentary tools, metal hammers, and machetes, to disassemble hazardous materials without adequate safety gear is not just a health risk but a structural issue, one that reflects the unregulated, underprotected conditions of informal e-waste economies (see also Akormedi et al., 2013; Lepawsky, 2015; Oteng-Ababio, 2012). This connects to broader scholarship on environmental injustice, which highlights how toxic labor is disproportionately offloaded onto poor, racialized, and gendered bodies in the global South (Nixon, 2011; Pellow, 2007). These labor conditions are symptomatic of a broader global division of environmental risk, where the burdens of technological progress, including the transition to so-called clean energy, are systematically offloaded onto marginalized populations in the global South, many of whom are women occupying the lowest rungs of economic hierarchies. The use of primitive tools in dismantling solar batteries is not due to ignorance or lack of innovation among workers; rather, it is the outcome of an economic architecture that rewards informality, displaces regulatory accountability, and renders the laboring body hyper-visible in its productivity but invisible in its suffering.

This invisibility is gendered in very specific ways. Women in e-waste economies, as in other low-wage economies, are often perceived as naturally suited to repetitive, meticulous tasks, making them ideal subjects for exploitative labor regimes that depend on low-cost, disposable workforces (Benería, 2001; Elson & Pearson, 1981; Sassen, 2000). Elson and Pearson's (1981) early analysis of women's industrial labor remains relevant here: what appears as skill or suitability is in fact the product of gendered expectations and historical devaluation, whereby women are inserted into global commodity chains precisely because their labor can be underpaid and overlooked. However, this 'suitability' is not an organic fact but a social construction, emerging from a long history of colonial and postcolonial gender regimes that position women, particularly poor women, as bearers of invisible, undervalued labor (Federici, 2012).

Within this configuration, women's bodies become the absorbers of risk, not just because they are cheap to employ, but because their injuries, exhaustion, and deaths seldom disrupt the flow of financial capital. From this perspective, the hammer and machete women use are not neutral tools, but proxies for an apparatus that disciplines through neglect, extracting the very materials of solar power capitalism through gendered and racialized disposability.

Additionally, the emotional exhaustion and the degrading treatment several women working in the e-waste site receive from their bosses bring to the fore the psychosocial toll of gendered subjugation within the so-called green labor regimes. The bodily injury, a severed toe from stepping on sharp metal, of the woman mentioned earlier is both a literal and symbolic wound. It serves as a reminder that beneath the surface of sustainability lies a brutal political economy of bodily risk, trauma, and abandonment. It is vital to see labor not only as economic but as embodied and affective (Fortunati, 2025; Gill & Pratt, 2008). This affective dimension of labor reveals how injury resonates beyond the body, destabilizing the intimate economies that sustain life.

Her inability to support her family during her recovery also highlights how workplace injury disrupts household stability and care, often plunging entire households into economic crisis. This aligns with the argument that the maintenance of life and household provisioning is not a private affair; it is intimately tied to, and often subordinated by, regimes of capitalist accumulation (Bhattacharya, 2017). The fact that her injury led to a complete cessation of her ability to support her family, without recourse to compensation, insurance, or state welfare, reveals how solar power capitalism is animated by logics of disposability. The worker is not viewed as a human being embedded in networks of care, responsibility, or community, but as a temporary utility. When an injury occurs, the cost is absorbed not by the employer or the system that profits from her labor, but by her own kin, community, and in this rare instance, strangers moved by compassion. This privatization of risk, wherein the consequences of structural violence are localized and depoliticized, ensures that the system remains intact or superficially altered, while the solar waste worker is left to absorb the fallout.

The employer's indifference to worker safety and wellbeing, the 'bosses don't care if I get injured,' is not simply a moral failure or an oversight; it is a feature of Achille Mbembe's (2020) necropolitics, the calculated exposure of certain populations to premature death in the service of economic life elsewhere. Her labor is embedded in this calculus, where bodily harm is normalized, even expected, as part of the job. The state's silence on such matters, its failure to regulate, protect, or even recognize this form of labor, further entrenches this necropolitical logic.

Her survival through the benevolence of good Samaritans rather than institutional support further illustrates the abandonment of laboring subjects by the state and other economic actors. This cannot be framed solely as failure of public health or occupational safety policy; it signals the operation of 'slow death' (Berlant, 2007), the wearing down of populations deemed economically useful but socially expendable. In this regime, women's ability to sustain families and communities is not simply neglected; the conditions of life and care are actively undermined by economic structures that extract labor while refusing to guarantee the conditions necessary for life.

And yet, her continued presence in this work, even after a life-altering injury, also points to the coercive nature of informal labor markets, where exit is rarely a real option. The threat of hunger, the pull of familial obligation, and the absence of alternatives force her to return to labor that is neither dignified nor safe. This is not resilience in the celebratory sense; it is survival under duress, enacted in the ruins of institutional care. Such women's labor thus is not only a material contribution to solar power capitalism; it attests to the forms of suffering that sustain its bright future.

Women's experiences of 'respect' or 'bad treatment' in the scrapyard also expose the social tensions arising from their expanded economic roles, which do not always conform to entrenched norms. Abina, a 43-year-old female boss, a rare leadership role for women in the scrapyard commented, 'Working at the scrapyard is seen as a man's job [...] you are discriminated against, especially if you are not doing it with your husband' (SWF4, Ghana, 2024). Her experience reveals the persistent regulation of women's economic activities through patriarchal family structures. The legitimacy of her income-generating work is mediated not by the quality or intensity of her labor, but by its perceived conformity to conventional domestic arrangements. This aligns with long-standing scholarship that women's entry into male-dominated economic spheres often provokes social backlash, especially when it disrupts dominant ideologies of gender complementarity and domestic respectability (Galea & Chappell, 2022; Martin & Barnard, 2013; Ong, 1987). The e-waste site becomes a contested space where normative femininity is both reasserted and renegotiated. Women's visibility in the informal e-waste economy, especially in leadership roles, subverts local gender hierarchies, but not without social cost. As the quote suggests, the absence of a male partner invites suspicion, ostracism, and discrimination, signaling that economic autonomy remains a fraught endeavor. This echoes Kabeer's (2002) observation that empowerment is not a linear process but one embedded in uneven negotiations with social norms, kinship structures, and institutional exclusions. Here, economic agency does not automatically translate into social empowerment; rather, it invites new forms of surveillance and moral judgment.

In contrast to Abina's experience of discrimination and surveillance, other women's narratives reveal how conformity to patriarchal arrangements can generate legitimacy and social insulation. On this point, Ajoba, a 32-year-old mother of five noted:

> I am a boss because of my experience, so my husband and I have boys who work for us. These workers help us earn enough income to meet our family's needs. We sell the scraps to other companies, some of which export them to manufacturing firms abroad. We specialize in solar scraps, so we receive them every day – we are well known for that. […] I have never been treated badly; I am a boss, and bosses are accorded the highest level of respect (WW16, Ghana, 2024).

Despite operating within the same labor-intensive and male-dominated sphere of the e-waste economy, her narrative reveals how embeddedness within normative family structures, particularly a working relationship with her husband, can insulate women from gendered stigma. Both Ajoba (aged 32) and Abina (aged 43) hold the title of 'boss' at the same e-waste site. In many African communities, age is closely tied to social respect and authority (Iyare et al., 2021; Moore & Puritt, 2017; Oppong, 2006). Given this socio-cultural norm, one would expect Abina, though not elderly, to command more respect than the younger Ajoba. However, the reverse appears to be the case. One possible explanation is that Ajoba's joint venture with her husband offsets age hierarchies, conferring legitimacy that Abina lacks. Ajoba's joint entrepreneurial venture with her spouse appears to confer not only material stability but also moral credibility. In contrast to Abina, who faced suspicion in the absence of a male partner, this woman's authority as a 'boss' is not questioned; rather, it is valorized. She states unequivocally, 'I have never been treated badly […] bosses are accorded the highest level of respect.' This divergence shows how gendered experiences of labor are mediated by familial configurations and the symbolic capital they produce. Her account suggests that patriarchal structures, while often constraining, can also enable, particularly when women's economic roles are framed as extensions of collective household enterprise rather than as individual pursuits of autonomy. In this respect, her respectability is not divorced from her gender role but anchored within it. The solar scrap business becomes an acceptable domain for entrepreneurial activity precisely because it does not overtly challenge the gendered moral economy of the household.

However, not all women are shielded by such familial configurations; for some, gender and ethnicity intersect to produce deeper forms of vulnerability. Single or younger women reported sexual harassment and ethnicized insults, as described by Afua, a 25-year-old woman who has been working at the e-waste site for two years:

> I am an Akan. The scrapyard workers are mostly Northerners. I am sometimes seen as a lazy woman who has come here to have sexual relations with Northern men. Some men insult me by using my tribe and speak to me disrespectfully. At times, some of them try to take advantage of me sexually. I reported a couple of them to my boss. They have not succeeded.

Such constrained contestation resonates with Kenyan accounts of silence as power, where women deploy nonconfrontational tactics to renegotiate hostile conditions without inviting reprisal (Gatwiri & Mumbi, 2016). Afua's experience reveals the layered vulnerabilities that women face at the intersection of gender, ethnicity, and labor hierarchies in the context of solar power capitalism. Her experience illustrates how structural inequalities are reproduced in the social organization of labor at e-waste sites. In such masculinized environments, some women's presence is interpreted through moralistic and ethnicized lenses. The labeling of Akan women as 'lazy' or 'sexually available' by predominantly Northern male workers points to the enduring cultural discourses that shape gendered labor relations. The experience of women like Afua echoes those of the Dalit women in India, whose public presence is often misconstrued as a sign of sexual availability, exposing them to sexual harassment (Gopal, 2013; John, 2013). These micro-level interactions reflect the overlapping systems of discrimination that cannot be understood separately from one another (see Crenshaw, 2013). Sexual harassment in such settings functions not only as a tool of individual dominance but also as a disciplinary mechanism that regulates women's access to economic participation. In intersectional terms, gender and ethnicity co-produce coerced compliance in the labor process. Informal labor markets, particularly in the global South, are rife with coercive gendered dynamics (Mezzadri, 2016). The boundaries of acceptable female behavior are policed not only by male workers but also through broader discourses of ethnic belonging, regional identity, and labor legitimacy. For women like Afua, merely showing up to work is itself a transgression, especially when their labor defies both traditional domestic expectations and the masculinized ethos of the e-waste site.

Despite this, within these constraining conditions, women like Afua carve out spaces of refusal and endurance. Afua's resistance, reporting harassment to her supervisor, should not be underestimated. This act reflects a form of embodied agency, where marginalized actors assert their rights within constraining structures (Jackson, 2007; Kawarazuka et al., 2019; McNay, 2013), often through subtle forms of refusal rather than overt protest.[6] Her assertion that 'they have not succeeded' is powerful; it signals not only her survival but also the small, yet vital, mechanisms through which women negotiate hostile labor spaces. In fact, Afua's actions disrupt the narrative that women in precarious labor sites are passive victims. Instead, she enacts a form of resistance, asserting boundaries within a hyper-masculinized and ethnically-stratified work environment. Her resistance is shaped by a constellation of social structures—patriarchy, ethnicity, class, and informal labor hierarchies—that simultaneously constrain and enable her agency. The experiences of women like Afua highlight the importance of analyzing solar power projects not only through technological or economic lenses but also through gendered labor and survival frameworks. Women's experiences in these spaces are shaped as much by transnational green energy actors as they are by local patriarchal orders. Women like Afua do not only labor with their bodies to dismantle non-functional solar devices, other electronics, and sort metals; they also labor emotionally and socially to survive spaces of hostility. The feminized body becomes both a site of exploitation and negotiation (Benya, 2015; Federici, 2004; Wright, 2013). While Afua's Akan identity places her in a minority position at the Agbogbloshie scrapyard, it also marks her as morally suspect in the eyes of male Northern coworkers, who draw on ethnonationalist imaginaries to police gendered behavior.

[6] Women's interpretations of their bodies, expressions of autonomy, and enactments of agency are shaped by complex, context-specific dynamics (Bawa & Adeniyi Ogunyankin, 2018; Oyěwùmí, 1997).

Moreover, Afua's experience foregrounds gendered survival work in the informal workplace, where women must continually reproduce their capacity to labor amid harassment, toxic exposures, and weak institutional protections. Without institutional protections, the burden of maintaining bodily integrity, emotional wellbeing, and personal safety falls on women themselves. This aligns with feminist framings of gendered survival as an ongoing, embodied struggle in which the conditions for life and labor are constantly produced and defended (Bakker, 2007; Federici, 2012; Ojeda, 2021). Seen through this lens, solar power capitalism is sustained not only by financial contracts or fenced land but by the exhaustion of women's bodies in toxic scrap economies, an ultimate form of insertion whose cumulative drain is depletion.

6.5 Concluding Remarks

Patriarchy is one of solar power capitalism's operating systems. Women's time, credit, and care labor are continually inserted into the circuitry of accumulation, and their exhaustion constitutes the unacknowledged depletion that keeps lights on and debts repaid. At the same time, within these burdens lie ambivalent gains, moments of relief, dignity, and incremental agency that complicate any simple narrative of victimhood. By revealing these contradictions, I have shown that insertion and depletion are gendered processes that render patriarchy infrastructural to green energy accumulation. With these dynamics established, I now turn in the concluding chapter to draw the threads together, linking extraction, finance, patriarchy, and toxic afterlives into a single circuit of solar power capitalism and asking what forms of repair or endurance might yet interrupt it.

References

Akese, G. A., & Little, P. C. (2018). Electronic waste and the environmental justice challenge in Agbogbloshie. *Environmental Justice, 11*(2), 77–83.

Akormedi, M., Asampong, E., & Fobil, J. N. (2013). Working conditions and environmental exposures among electronic waste workers in Ghana. *International Journal of Occupational and Environmental Health, 19*(4), 278–286.

Amone, C., Bond, J., Chiweshe, M. K., Fusari, V., Malunga, B., Mandhu, F., Ndi, C. L. A. E., Obiora, L. A., Tandrayen-Ragoobur, V., Seye, C. A., & Siang'andu, E. T. (2021). *Patriarchy and gender in Africa.* Rowman & Littlefield.

Bakker, I. (2007). Social reproduction and the constitution of a gendered political economy. *New Political Economy, 12*(4), 541–556.

Barca, S. (2020). *Forces of reproduction: Notes for a counter-hegemonic Anthropocene*. Cambridge University Press.

Bawa, S. (2016). Paradoxes of (dis)empowerment in the postcolony: Women, culture and social capital in Ghana. *Third World Quarterly, 37*(1), 119–135.

Bawa, S. (2012). Women's rights and culture in Africa: A dialogue with global patriarchal traditions. *Canadian Journal of Development Studies/Revue canadienne d'études du Développement, 33*(1), 90–105.

Bawa, S., & Adeniyi Ogunyankin, G. (2018). (Un) African women: Identity, class and moral geographies in postcolonial times. *African Identities, 16*(4), 444–459.

Benería, L. (2001). Shifting the risk: New employment patterns, informalization, and women's work. *International Journal of Politics, Culture, and Society, 15*, 27–53.

Benya, A. (2015). The invisible hands: Women in Marikana. *Review of African Political Economy, 42*(146), 545–560.

Berlant, L. (2007). Slow death (sovereignty, obesity, lateral agency). *Critical Inquiry, 33*(4), 754–780.

Bhandar, B. (2018). *Colonial lives of property: Law, land, and racial regimes of ownership.* Global and Insurgent Legalities. Duke University Press.

Bhattacharya, T. (Ed.). (2017). *Social reproduction theory: Remapping class, recentering oppression.* Pluto Press.

Bhattacharyya, G. (2018). *Rethinking racial capitalism: Questions of reproduction and survival*. Rowman & Littlefield.

Bouzarovski, S., & Petrova, S. (2015). A global perspective on domestic energy deprivation: Overcoming the energy poverty–fuel poverty binary. *Energy Research & Social Science, 10*, 31–40.

Crenshaw, K. W. (2013). Mapping the margins: Intersectionality, identity politics, and violence against women of color. In *The public nature of private violence* (pp. 93–118). Routledge.

Dankwa, S. (2018). Culture of family ideals and perceived subjugating positions of women in patriarchy society: The way forward. *Culture (Québec), 8*(24).

Dogo, S. A. (2014). The Nigerian patriarchy: When and how. *Cultural and Religious Studies, 2*(5), 263–275.

Elson, D., & Pearson, R. (1981). 'Nimble fingers make cheap workers': An analysis of women's employment in third world export manufacturing. *Feminist Review, 7*(1), 87–107.

Federici, S. (2004). *Caliban and the witch: Women, the body, and primitive accumulation.* Autonomedia.

Federici, S. (2012). *Revolution at point zero: Housework, reproduction, and feminist struggle.* PM Press.

Federici, S. (2020). *Revolution at point zero: Housework, reproduction, and feminist struggle.* PM Press.

Fernandes, M., Lupo, L., Benya, A., Dedeoğlu, S., Mezzadri, A., & Prügl, E. (2023). Social reproduction, women's labour and systems of life: A conversation. *Dialogues in Human Geography, 13*(3), 473–483.

Fortunati, L. (2025). *The arcana of reproduction: Housewives, prostitutes, workers and capital.* Verso Books.

Fraser, N. (2016). Contradictions of capital and care. *New Left Review, 100,* 99–117. https://newleftreview.org/issues/ii100/articles/nancy-fraser-contradictions-of-capital-and-care

Galea, N., & Chappell, L. (2022). Male-dominated workplaces and the power of masculine privilege: A comparison of the Australian political and construction sectors. *Gender, Work & Organization, 29*(5), 1692–1711.

Gatwiri, G. J., & Mumbi, K. A. (2016). Silence as power: Women bargaining with patriarchy in Kenya. *Social Alternatives, 35*(1), 13–18.

Gill, R., & Pratt, A. (2008). In the social factory? Immaterial labour, precariousness and cultural work. *Theory, Culture & Society, 25*(7–8), 1–30.

Gopal, M. (2013). Ruptures and reproduction in caste/gender/labor. *Economic and Political Weekly, 47*(18), 91–97.

Hochschild, A., & Machung, A. (2012). *The second shift: Working families and the revolution at home.* Penguin.

Hunnicutt, G. (2009). Varieties of patriarchy and violence against women: Resurrecting "patriarchy" as a theoretical tool. *Violence against Women, 15*(5), 553–573.

Iyare, A. E., Imafidon, E., & Abudu, K. U. (2021). Ageing, ageism, cultural representations of the elderly and the duty to care in African traditions.

Essays on contemporary issues in African philosophy (pp. 281–299). Springer International Publishing.
Jackson, C. (2007). Resolving risk? Marriage and creative conjugality. *Development and Change, 38*(1), 107–129.
John, M. (2013). The problem of women's labor: Some autobiographical perspectives. *Indian Journal of Gender Studies, 20*(2), 177–212.
Kabeer, N. (2002). *The power to choose: Bangladeshi women and labor market decisions in London and Dhaka.* Verso.
Kandiyoti, D. (1988). Bargaining with patriarchy. *Gender & Society, 2*(3), 274–290.
Kaushal, T. (2025). Gendered labour produces finance: Navigating profits and care as intermediaries of finance. *Economy and Society*, 1–24.
Kaushal, T., & Predmore, S. (2025). The women's inclusion agenda: Gender and everyday practices across registers of finance. *Finance and Society, 11*(1), 81–102.
Kawarazuka, N., Locke, C., & Seeley, J. (2019). Women bargaining with patriarchy in coastal Kenya: Contradictions, creative agency and food provisioning. *Gender, Place & Culture, 26*(3), 384–404.
Khandker, S. R., Samad, H. A., Sadeque, Z. K., Asaduzzaman, M., Yunus, M., & Haque, A. E. (2014). *Surge in solar-powered homes: Experience in off-grid rural Bangladesh.* World Bank Publications.
Klingler, M., Ameli, N., Rickman, J., & Schmidt, J. (2024). Large-scale green grabbing for wind and solar photovoltaic development in Brazil. *Nature Sustainability, 7*(6), 747–757.
Lemaire, X. (2018). Solar home systems and solar lanterns in rural areas of the Global South: What impact? *Wiley Interdisciplinary Reviews: Energy and Environment, 7*(5), Article e301.
Lepawsky, J. (2015). The changing geography of global trade in electronic discards: Time to rethink the e-waste problem. *The Geographical Journal, 181*(2), 147–159.
Lingham, J. T., & Johnston, M. (2025). Running on empty: Depletion and social reproduction in Myanmar and Sri Lanka. *Antipode, 57*(2), 494–514.
Magaramombe, G. (2010). 'Displaced in place': Agrarian displacements, replacements and resettlement among farm workers in Mazowe District. *Journal of Southern African Studies, 36*(2), 361–375.
Martin, P., & Barnard, A. (2013). The experience of women in male-dominated occupations: A constructivist grounded theory inquiry. *SA Journal of Industrial Psychology, 39*(2), 1–12.
Mbembe, A. (2020). *Necropolitics.* Duke University Press.

McNay, L. (2013). *Gender and agency: Reconfiguring the subject in feminist and social theory.* Wiley.
Mezzadri, A. (2016). *The sweatshop regime: Labouring bodies, exploitation, and garments made in India.* Cambridge University Press.
Mollett, S. (2014). A modern paradise: Garifuna land, labor, and displacement-in-place. *Latin American Perspectives, 41*(6), 27–45.
Moore, S. F., & Puritt, P. (2017). *The Chagga and Meru of Tanzania: East Central Africa Part XVIII.* Routledge.
Murrey, A. (2015). Narratives of life and violence along the Chad-Cameroon Oil Pipeline. *Human Geography, 8*(1), 15–39.
Nixon, R. (2011). *Slow violence and the environmentalism of the poor*. Harvard University Press.
Ochieng' Obado, E. A., Odera, J. A., & Sindiga, I. (1995). *Management of medicinal plant resources in Nyanza, traditional medicine in Africa.* East African Educational Publishers Ltd.
Odore, H. C. (2002). *Indigenous knowledge and the integration of knowledge systems – Towards a philosophy of articulation.* New African Books.
Ojeda, D. (2021). Social reproduction, dispossession, and the gendered workings of agrarian extractivism in Colombia. In *Agrarian extractivism in Latin America* (pp. 85–98). Routledge.
Omstedt, M., & Ebner, N. (2024). Introduction: Uneven development and social difference in capitalism. *Environment and Planning A: Economy and Space, 56*(4), 1298–1303.
Ong, A. (1987). *Spirits of resistance and capitalist discipline: Factory women in Malaysia.* SUNY Press.
Oppong, C. (2006). Familial roles and social transformations: Older men and women in sub-Saharan Africa. *Research on Aging, 28*(6), 654–668.
Ossome, L. (2024). Imperialism and crises of social reproduction in Africa. *Review of African Political Economy, 51*(181), 515–530.
Oteng-Ababio, M. (2012). When necessity begets ingenuity: E-waste scavenging as a livelihood strategy in Accra, Ghana. *African Studies Quarterly, 13.*
Ouma, A. (2022). Intergenerational learning processes of traditional medicinal knowledge and socio-spatial transformation dynamics. *Frontiers in Sociology, 7*, Article 661992.
Oyěwùmí, O. (1997). *The invention of women: Making an African sense of Western gender discourses.* University of Minnesota Press.
Pellow, D. N. (2007). *Resisting global toxics: Transnational movements for environmental justice.* MIT Press.

Rai, S. M., Hoskyns, C., & Thomas, D. (2014). Depletion: The cost of social reproduction. *International Feminist Journal of Politics, 16*(1), 86–105.
Sanya, B. N. (2013). Disrupting patriarchy: An examination of the role of e-technologies in rural Kenya. *Feminist Africa, 18*, 12–24.
Sassen, S. (2000). Women's burden: Counter-geographies of globalization and the feminization of survival. *Journal of International Affairs,* 503–524.
Standal, K., & Winther, T. (2016). Empowerment through energy? Impact of electricity on care work practices and gender relations. *Forum for Development Studies, 43*(1), 27–45.
Stock, R., & Birkenholtz, T. (2024). Tracking the sun: Exposing India's solar dispossessions. *Globalizations,* 1–20.
Vergès, F. (2021). *A decolonial feminism.* Pluto Press.
Wright, M. (2013). *Disposable women and other myths of global capitalism.* Routledge.
Yenneti, K., Day, R., & Golubchikov, O. (2016). Spatial justice and the land politics of renewables: Dispossessing vulnerable communities through solar energy mega-projects. *Geoforum, 76*, 90–99.

7

Circuits of Exhaustion

7.1 Confronting Contradictions

Solar power is widely celebrated as the emblem of a green future, a technological fix that promises to end energy poverty, particularly in the global South, while cutting carbon. But beneath this optimism lies a deeper tension, one that resonates across global capitalism. This tension crystallizes most clearly through insertion and depletion: incorporation into circuits of accumulation, and simultaneous exhaustion to stabilize those circuits. Like the specter Marx once evoked, solar power haunts our imaginaries, appearing as both deliverance and menace. It brightens homes, but it also obscures the violence of mines, the enclosures of land, the burdens of debt, and the afterlives of toxic solar waste. To ask whether solar power is a solution is therefore to ask: a solution for whom, under what conditions, and at what cost?

Throughout this book I have argued that we can only answer this question by mobilizing a conceptual apparatus attuned to the contradictions of global capitalism: uneven development, financialization, patriarchy, insertion, and depletion. These are lenses sharpened through empirical encounters. From the bauxite dust of Awaso to the fences of Kaleo-Lawra (Ghana), from the household debts of women in Kenya,

N. Ojong, *Solar Power Capitalism*, Energy, Climate and the Environment,
https://doi.org/10.1007/978-3-032-13707-4_7

Tanzania, Malawi, and Cameroon to the smoldering remains of Agbogbloshie (Ghana), each case exposes how solar power capitalism organizes value by shifting costs onto poorer households and peripheral ecologies. What anchors these sites is not only 'insertion' but also 'depletion': the steady draining of marginalized people's time and bodies, much like depleted soils and polluted rivers, to stabilize accumulation. These downward and outward shifts are not gender-neutral. They are routed through patriarchal governance, which positions women as the guarantors of repayment, the absorbers of risk, and the caretakers of households stretched by enclosure and toxicity. This dynamic exemplifies a broader global pattern in which, as Kaushal and Predmore (2025) argue, women's 'inclusion' serves to stabilize markets while shifting the costs of risk downstream. Patriarchy, in this sense, is infrastructural: a social architecture of insertion that enables accumulation while naturalizing the transfer of burdens from global investors onto women's labor, reputations, and social standing. Here too, patriarchy is best understood as a modality of insertion and depletion, structuring how incorporation and exhaustion are gendered.

Methodologically, I treat insertion and depletion as mutually-constitutive moments of a single circuit. Insertion—the incorporation of people, ecologies, and institutions into infrastructures of accumulation—produces the very exhaustion I frame as depletion. Depletion, in turn, deepens the conditions for further insertion, as scarcity of time, health, and ecological resources compels renewed engagement with solar finance, enclosure, and solar waste. Each process therefore produces and depends on the other, a dialectic that underwrites the stability of solar power capitalism itself. As the book's organizing framework, insertion and depletion provide a distinct conceptual architecture that renders visible disparate terrains through one conceptual lens. In other words, both concepts ensure that bauxite mines, fenced commons, mobile-money repayment platforms, and toxic solar waste are grasped as variations of the same circuitry. Table 7.1 illustrates how each empirical site can be read as a knot in the circuit of insertion and depletion—never as isolated harm, but as systematic incorporation and exhaustion. It reveals how every empirical site analyzed in this book operates as a knot in the circuit of insertion and depletion.

Table 7.1 Sites, mechanisms, and burdens in solar power capitalism

Site	Mechanism of accumulation	Burdens/Consequences
Awaso (Ghana)	Bauxite extraction feeding global solar supply chains	Bauxite dust-laden lungs, degraded soil, precarious mine labor
Kaleo-Lawra (Ghana)	Utility-scale solar plant on communal land, enclosure of commons	Displacement, rerouted women's labor, time tax on households
Agbogbloshie (Ghana)	Informal recycling of discarded solar goods, toxic value recovery	Burns, poisoned soils, shortened lifespans, necropolitical abandonment
Kenya, Tanzania, Malawi, Cameroon	PAYG solar home systems, securitization of micro-debts, financial inclusion logics	Household indebtedness, energy precarity, erosion of dignity, gendered depletion

The conceptual apparatus guiding this investigation rests on four interlocking strands. These are not independent pillars but supporting scaffolds. Each helps sharpen the visibility of insertion and depletion, which remain the core theoretical lens.

Uneven development clarifies where insertion takes place; financialization shows how insertion is capitalized; structural violence and necropolitics expose how depletion is normalized. More precisely, uneven development clarifies how solar energy expansion reproduces colonial logics of selective provision (see also Omstedt & Ebner, 2024; Ortiz, 2025; Smith, 2008), shifting capital to new frontiers while leaving costs behind. Financialization demonstrates how kilowatts become cash flows, as mobile-money transfers in Accra or Kisumu are bundled into securitized assets marketed in major financial centers (see also Birch, 2024; Bracking, 2020; Roitman, 2025). The securitization of mobile-money repayments is thus not just financialization but a form of data-value production (Roitman et al., 2025), where the digital imprints of everyday energy use are capitalized as both speculative and operational value. Structural violence (Farmer, 2004; Galtung, 1969) exposes how such abstractions materialize as shortened life chances, whether through bauxite dust-laden lungs in Awaso or the time lost in reroutes around the Kaleo-Lawra solar plant. By centering women's insertion, I have shown how the reliability of the system rests on the conversion of women's temporal, financial, and corporeal work into the infrastructure of solar energy markets. Alongside these, I have theorized depletion (Lingham & Johnston, 2025; Rai et al., 2014), not as a sign of policy failure but as a mechanism through which green value is secured.

Across the chapters, I have highlighted a single circuit linking extraction, enclosure, household finance, and toxic afterlives, a chain of accumulation whose details unfold in the next section. Table 7.1 distills these sites, mechanisms, and burdens, providing a synoptic guide to the empirical trajectory developed throughout the book.

The sites covered in this book make visible more than the contradictions of one energy sector. They reveal how contemporary capitalism governs through patterns of insertion and depletion, how bodies, ecologies, and financial infrastructures are reconfigured to stabilize accumulation. This book thus contributes not only to debates on energy justice but to the broader theorization of global capitalism. Insertion and depletion offer a conceptual lens for apprehending how capital's latest fixes, whether solar plants in Ghana, lithium mines in Bolivia, or fintech infrastructures in India, depend on the downward shifting of costs and the outward extension of risk.

Across the cases, the coherence of solar power capitalism rests not only on insertion and depletion but also on the three design logics introduced in Chapter 2—enclave privilege, externalization of risk, and concessional underwriting—which persist as the infrastructural blueprints through which accumulation is organized and legitimated.

The sections that follow develop this argument: Sect. 7.2 offers a close empirical journey through Ghana's mine, solar plant, household, and scrapyard; Sect. 7.3 dwells on the ambivalent gains and losses that women and households articulate; Sect. 7.4 then widens the lens to Latin America and India to situate Africa's solar energy frontier within global green capitalism; and Sect. 7.5 closes by sketching the fragile but vital alternative futures already taking shape in everyday practices of care and solidarity.

7.2 Reworking the Circuit

To grasp the contours of solar power capitalism, one must hold together the multiple spaces where it takes shape. As discussed in the previous section, the dynamics of insertion and depletion, routed through patriarchal governance, bind these sites together.

At the point of extraction, Awaso embodies how solar energy's supply chain begins in the mine. Roadside vendors inhale bauxite dust, contract workers rotate through precarious shifts, and corporate social responsibility schemes remain threadbare. This is not collateral damage but accumulation by dispossession (Harvey, 2003): landscapes torn apart so that surplus value can be realized elsewhere. But the mine is more than a local rupture; it is a conduit in the wider geography of uneven development. This is precisely the 'seesaw' dynamic Smith (2008) described in which capital migrates in search of higher returns, leaving behind landscapes of dust and dispossession. What seems as a process of raw material supply for solar goods is also a logic of capital's restless mobility, the competition between places to host investment, and the reproduction of underdevelopment in its wake. In Awaso, the global appetite for green minerals is lived as bauxite dust and depleted soil, a reminder that the promise of clean energy elsewhere is anchored to the pollution and precariousness of the periphery.

At Kaleo and Lawra, solar power capitalism takes the form of enclosure. Fences cordon off land that once supported grazing and farming, rendering livelihoods precarious under the guise of renewable energy. Compensation schemes are uneven while the costs of exclusion are immediate and tangible. Enclosure here is not simply a matter of land boundaries; it is 'climate necropolitics' (Deberdt and Le Billon, 2024: 1264)—the reorganization of life chances along carbon hierarchies.[1] Those with access to climate capital, able to invest in its infrastructure, are positioned as beneficiaries, while those dispossessed of land are relegated to classes that remain carbon-dependent or excluded from transition altogether. From this lens, beyond being an energy project, the

[1] For a detailed discussion of 'climate necropolitics,' see DeBoom (2021).

solar plant is a sorting mechanism, distributing privilege and vulnerability in ways that mirror older colonial geographies of extraction. What is framed as a green transition is also a violent transition, one that redraws the boundaries of survival itself.

The circuit then extends into the household, where finance mediates energy access. PAYG SHS rent-to-own contracts demand regular installments, with missed payments leading to immediate disconnection. Families stretch repayment schedules through kin loans, divert remittances, or borrow from indigenous financial institutions to keep the lights on. What is framed as flexibility, in the form of small incremental installments, quickly reveals itself as fragility, where a single bad harvest or illness can extinguish light and erode social standing. This financial yoking of households materializes the 'financialization of life' (Langley, 2020): debt and repayment schedules seep into the routines of care, eating, and sleeping. Access to renewable energy here operates as a governance of life itself. Finance does not only bind households to distant investors; it remakes the ecological frontier, as Bracking (2024) shows, by converting everyday practices of subsistence into securitized revenue streams. The SHS, often framed as empowerment (Khasru & Rafee, 2021; Winther et al., 2018), hence folds domestic survival into global circuits of accumulation, where the price of illumination is paid in both cash and exhaustion.

This dynamic of household finance is inseparable from gendered relations of obligation and control. While PAYG finance reorganizes household life, it does so most directly through women. Across Kenya, Tanzania, Malawi, and Cameroon, women are the ones who often shoulder repayment, absorb the shame of disconnection, and negotiate with lenders or indigenous savings and credit institutions when money falls short. Their insertion into solar energy capitalism is constitutive of its very functioning.

Uneven development is always articulated through social difference: gender, class, and race are not external to accumulation but the very grounds on which it operates (Omstedt & Ebner, 2024). Women's time, labor, and reputations often become the infrastructure that stabilizes returns in volatile economies. Framing this solely as household adaptation would be to miss the deeper structure at work. Hunnicutt (2009) contends that patriarchy endures as a systemic logic of domination, shaping institutions and social orders beyond the intimate sphere. The reliance on women to manage solar debt reproduces this: patriarchal governance not only devalues women's claims to security and recognition but simultaneously harnesses their agency as a stabilizing force. What could be considered as energy inclusion is another iteration of patriarchal capitalism, where women's bodies and creditworthiness secure value for others while protections for them remain meager.

The circuit's final movement unfolds at the end of the product life cycle. Agbogbloshie in Accra speaks to the toxic legacies of solar goods. Discarded solar panels, batteries, etc. are broken down with rudimentary tools in open-air sites where smoke and chemical residues hang thick in the air. This constitutes what Murrey and Mollett (2023) describe as 'extractive logics': racialized and gendered violence written onto bodies and ecologies.

In Agbogbloshie, dismantling wires, batteries, and circuit boards is a continuation of extraction by other means, where e-waste itself becomes profitable. Women and children labor in these spaces of slow violence, their embodied depletion folded into the very reproduction of solar energy capitalism's global circuitry.

The very logics of disposability are gendered and racialized, as the bodies of Black women and children (as well as other marginalized bodies) are rendered expendable in ways that reinscribe colonial hierarchies of waste and worth. Hazardous residues here are not by-products of poor governance but organized forms of value extraction, where health, reproduction, and survival are systematically traded off for meager livelihoods. Recycling is another frontier of dispossession, where survival is pieced together through the inhalation of toxic fumes and the poisoning of lungs, landscapes, and futures. Toxic residues mark Black bodies and ecologies as disposable, enacting what Black ecological scholarship describes as a racial capitalist ordering of life and death (Bruno, 2024; Hosbey et al., 2022). From this angle, the toxic afterlife of solar power capitalism constitutes a systematic production of disposability through the entwined operations of extraction, race, and gender.

As noted previously, these sites of extraction at Awaso, enclosure at Kaleo-Lawra solar plant, household finance, and toxic afterlife at

Agbogbloshie do not stand apart as isolated moments. They form an interlocked accumulation cycle in which value is realized through systematic displacement. Bauxite dust and dispossession, fences, debts, fumes, wasted bodies, and foreclosed futures: each is a register of how solar power capitalism organizes life. To see them in relation is to grasp not only how green energy is produced but also how its costs are distributed—downward onto the poor and outward to the periphery. This is the hidden architecture of the solar energy transition, an architecture built less on sunshine than on sacrifice.

7.3 The Contradiction of Ambivalent Gains

One of the most striking features of *Solar Power Capitalism* is how often women themselves insist on the tangible benefits of solar energy. They speak of relief from kerosene smoke, the pride of a well-lit home, and the joy of seeing children study after dusk (see Chapter 6). In Kaleo, some women described how electric lighting extended their capacity to cook or complete chores at night. In Lawra, some women turned chilled drinks into petty commerce. In Tanzania, a mother explained that solar freed her family from the fire hazard of kerosene lamps and allowed her children to study without risk to their eyes. These are not trivial observations. They cut to questions of health, dignity, and the aspirations households hold for the next generation.

However, the very conditions that enable these improvements also generate new forms of dependence and constraint. The same solar system that eradicates smoke also binds households to rolling micro-debts, with the threat of disconnection always close at hand for those with PAYG rent-to-own contracts with outstanding balances. The same solar plant that powers Kaleo and Lawra communities adds miles to women's daily journeys for water or fuelwood. In Accra, the afterlife of those solar technologies appears in the fumes of Agbogbloshie, where disposable workers inhale toxic smoke while salvaging discarded systems. Relief in one register returns as depletion in another.

Holding these contradictions together demands an analytic approach that neither romanticizes empowerment nor flattens it into exploitation.

It would be misleading to collapse these accounts into a single narrative of exploitation. Doing so would erase women's appreciation of what solar energy makes possible. But it would be equally naive to accept empowerment at face value. Their experiences point instead to the Marxist contradiction at the heart of capitalism: the very process that generates new capacities simultaneously deepens subordination (Fraser, 2016; Marx & Engels, 1848).

Ambivalence is the analytic ground of analysis. The double character of solar energy—its capacity to improve life while entangling it more tightly to patriarchy, debt, and enclosure—defines solar power capitalism. It is this simultaneity of relief and burden, of light and depletion, that makes solar energy such a potent frontier of accumulation. Framing solar energy only as promise would misrepresent the data; representing it only as harm would silence the women who insist on the dignity and social standing that light brings. The challenge is to hold both together, as contradiction. These ambivalences, however, are not confined to individual households; they resonate outward, linking Africa's solar energy frontier to wider debates on energy justice and the trajectories of global capitalism.

7.4 From African Frontiers to Global Capitalism

The empirical cases examined across Ghana, Kenya, Tanzania, Malawi, and Cameroon are not localized cases. They speak directly to broader debates on energy justice (Jenkins et al., 2021; Lacey-Barnacle et al., 2020; McCauley et al., 2013; Sovacool et al., 2020), climate finance (Ciplet et al., 2022; Dafermos, 2025; Navarro, 2022), and the trajectories of global capitalism (Ricci, 2021; Smith, 2016; Soederberg, 2020; Vidal, 2019). By following the circuits of solar power capitalism from mine to household and scrapyard, this book situates Africa not at the margins of the green transition but at its center. The continent is both a laboratory where new financial instruments are experimented with and a sacrifice zone where the residues of those experiments accumulate.

This positioning complicates conventional framings of energy justice that equate progress with access alone. Energy justice debates often focus on distribution, asking who gains access, at what price, and with what recognition of difference (Jenkins et al., 2016; Ojong, 2022; Sovacool & Dworkin, 2014). I pushed the conversation further. In Kaleo and Lawra, electricity is indeed distributed, but at the cost of longer walks for women and the enclosure of grazing land. In Kenya and Tanzania, access is achieved, but only through precarious repayment ecologies that discipline households to prioritize PAYG SHS installments over meals and other essential needs. At Agbogbloshie, solar goods once celebrated as a panacea to climate change reappear as toxic waste handled by precarious workers. Energy justice, in these contexts, cannot be reduced to access alone; it must account for the structural conditions under which that access is secured, and the bodies and ecologies that bear its hidden costs.

The same logic that governs solar energy access also organizes climate finance. What emerges is a vision of climate finance less about decarbonization than about asset creation. SHSs, securitized repayment streams, and blended-finance vehicles are valorized as sustainable innovations, but their design reveals how deeply the green energy transition is entwined with speculative accumulation. This is why the promise of a just energy transition rings hollow when viewed from Awaso's bauxite dust-laden streets or from the vantage of women choosing between food and light. Taking energy justice seriously requires asking not only who receives solar power but also who absorbs the costs of making that electricity investable.

These dynamics within Africa resonate with other geographies. As in Soederberg (2020), who analyzes housing struggles in Berlin, Vienna, and Dublin to reveal how surplus populations are managed in Europe, *Solar Power Capitalism* demonstrates how surplus capital finds new life in Africa's solar energy frontier. The logics are strikingly similar: in both cases, finance resolves its crises by displacing risk downstream and outward, producing new vulnerabilities while claiming to expand access.

The sites in Africa mirror global green frontiers shaped by similar extractive logics. In Latin America's lithium triangle, water tables are drained (Jerez et al., 2021; Kingsbury, 2023; Voskoboynik & Andreucci, 2022), so that batteries for electric vehicles can be produced.

In India, utility-scale solar plants displace agrarian livelihoods and reconfigure ecologies (Stock & Birkenholtz, 2024; Stock & Sovacool, 2023), all under the banner of clean energy and national energy security agendas. These projects differ in ecology and politics, but they share Africa's underlying calculus: accumulation through depletion, and the outsourcing of risk and toxicity to racialized and gendered peripheries. Africa's solar frontier is thus emblematic, not exceptional, within global green capitalism.

Recognizing this comparative horizon helps clarify what is at stake for global capitalism itself. Solar power capitalism is not solely about Africa's future; it is about the shape of global capitalism in an era of climate crisis. If the transition to renewable energy sources proceeds through the same logics of uneven development, financialization, and patriarchal governance demonstrated here, then the transition will replicate, rather than resolve, entrenched inequalities. The lesson is sobering but clear: decarbonization, if built on the foundations of fossil capitalism (see Chapter 3), will not deliver justice.

These parallels point to the central contradiction that opened this chapter. A green transition routed through the very infrastructures of fossil capitalism, with its uneven development, financial speculation, and patriarchal governance, cannot be just. Whether in the lithium sites of South America, the solar plants of India, or the fenced landscapes of Ghana, decarbonization pursued through these logics recasts extraction and depletion as progress, confirming that the promise of clean energy will remain haunted by the violence it seeks to transcend.

The dynamics I have demonstrated across these sites reaffirm the structural blueprint first outlined in Chapter 2: enclave privilege, risk externalization, and concessional underwriting. These design logics remain the enduring architecture through which solar power capitalism organizes life and value. Enclave privilege governs who gains access and whose lands are fenced; externalization of risk ensures that financial, ecological, and bodily costs are displaced onto households and women; and concessional underwriting secures legitimacy for these arrangements under the rubric of development. Collectively, they complete the circuit, reproduced from colonial grids to contemporary solar finance and thereby closing the analytic cycle of insertion and depletion that runs through this book.

7.5 Toward Alternative Futures

The argument developed in this book is sobering but it is not meant to paralyze. The purpose of insisting on contradiction, harm, and depletion is not to extinguish hope but to clarify where genuine alternatives might take root. Solar power capitalism reveals much about the limits of market-led transition, but it also makes visible the creativity and resilience of those compelled to navigate its terrain.

The expressions of alternative energy politics emerge not in grand policy frameworks but in the ordinary practices of survival and care. As noted previously in this book, some women in Kenya borrow strategically from indigenous savings and credit institutions, sometimes delaying installments to free up cash for school fees, knowing that the first PAYG SHS lockout is temporary and reversible. In Malawi, some farmers pool seasonal earnings to purchase solar panels outright, sidestepping the trap of rolling micro-debts. In Tanzania, households stretch the reach of light by sharing with neighbors, expanding access through solidarities invisible to investors. In Cameroon, households improvise to keep their SHSs running despite limited access to technicians and costly parts. Even at Agbogbloshie (Ghana), where toxic smoke fills the air, workers have organized informal systems of cooperation and knowledge-sharing that enable them to endure the hazards of dismantling e-waste. These practices are not blueprints for a just energy future. Nevertheless, they reveal how everyday agents rework the contours of solar power capitalism in ways that sustain life against its constraints.

These practices do not stand outside insertion and depletion; they rework them from below. What was once extraction becomes partial re-appropriation—an effort to bend the circuitry toward sustenance rather than exhaustion.

The challenge, then, is to imagine structural change that draws from these practices while moving beyond their limits. It is vital to imagine an energy transition organized not around collateral and securitization but around care and solidarity (see Magnani & Osti, 2016; Pellicer-Sifres, 2020; Sareen et al., 2023; Seyfang & Haxeltine, 2012). What would it mean to design infrastructures that honor the rhythms of daily reproduction over the demands of distant investors? What would energy systems look like if they were treated as commons to be defended rather than assets to be traded? These are not abstract theoretical exercises. They arise directly from the struggles of women calculating PAYG SHS installments against meals, farmers rerouted around fenced-off solar plants, and workers inhaling fumes as they strip discarded solar systems.

To insist on alternatives is not to romanticize improvisation or overlook the weight of structural constraints. It is, rather, to recognize that the seeds of a different energy politics are already visible in the cracks of solar power capitalism itself. Indigenous savings and credit institutions that buffer repayment crises, informal sharing of energy across households, and other grassroots initiatives gesture toward another vision, one where energy is organized around collective survival rather than financial return. This is not resistance in the romantic sense; it is repair work under constraint. Insertion and depletion remain the infrastructure of survival, but within them flicker other logics: reciprocity, redistribution, refusal.

Whether that vision can be realized is uncertain. What is certain, however, is that the present trajectory, where solar energy is financialized, solar plants are fenced, and broken solar systems are discarded into toxic waste sites, cannot be mistaken for justice.

I have shown that the contradictions of solar power are not peripheral flaws but the very logic of accumulation itself. Confronting this reality is unsettling, but necessary: it demonstrates that what passes as a green energy transition is in fact another round of depletion, another regime of patriarchal extraction framed as promise. And yet, within the cracks of that system, marginalized people's improvisations, solidarities, and acts of endurance point to another vision, one where energy might be organized around care, commons, and collective survival.

To conflate depletion with progress is to dim the future; to acknowledge it as such is the first step toward illuminating another path. The contradictions of solar power capitalism will continue to haunt any green transition that mistakes depletion for progress. This book has uncovered the circuitry of insertion and depletion; it closes by asking how that circuitry might be short-circuited, how the infrastructures of exhaustion could be repurposed for collective endurance.

The analysis in this book reveals that solar power capitalism does not only involve extraction or finance; it endures because insertion and depletion form the twin engines of its reproduction, entangling peripheral lives and ecologies to the stability of global green value.

References

Birch, K. (2024). Assetization as a mode of techno-economic governance: Knowledge, education and personal data in the UN's System of National Accounts. *Economy and Society, 53*(1), 15–38.

Bracking, S. (2020). Financialization and the environmental frontier. In *The Routledge international handbook of financialization* (pp. 213–223). Routledge.

Bracking, S. (2024). Green bond market practices: exploring the moral 'balance' of environmental and financial values. *Journal of Cultural Economy, 17*(3), 279–296.

Bruno, T. (2024). Black ecological relations and methodologies. *Progress in Environmental Geography, 3*(3), 271–285.

Ciplet, D., Falzon, D., Uri, I., Robinson, S. A., Weikmans, R., & Roberts, J. T. (2022). The unequal geographies of climate finance: Climate injustice and dependency in the world system. *Political Geography, 99*, Article 102769.

Dafermos, Y. (2025). Climate finance and global justice. *Climate Policy*, 1–17.

Deberdt, R., & Le Billon, P. (2024). Green transition's necropolitics: Inequalities, climate extractivism, and carbon classes. *Antipode, 56*(4), 1264–1288.

DeBoom, M. J. (2021). Climate necropolitics: Ecological civilization and the distributive geographies of extractive violence in the Anthropocene. *Annals of the American Association of Geographers, 111*(3), 900–911.

Farmer, P. (2004). An anthropology of structural violence. *Current Anthropology, 45*(3), 305–325.

Fraser, N. (2016). Contradictions of capital and care. *New Left Review, 100,* 99–117. https://newleftreview.org/issues/ii100/articles/nancy-fraser-contradictions-of-capital-and-care

Galtung, J. (1969). Violence, peace, and peace research. *Journal of Peace Research, 6*(3), 167–191.

Harvey, D. (2003). *The new imperialism.* Oxford University Press.

Hosbey, J., Lloréns, H., & Roane, J. T. (2022). Introduction: Global black ecologies. *Environment and Society, 13*(1), 1–10.

Hunnicutt, G. (2009). Varieties of patriarchy and violence against women: Resurrecting "patriarchy" as a theoretical tool. *Violence against Women, 15*(5), 553–573.

Jenkins, K., McCauley, D., Heffron, R., Stephan, H., & Rehner, R. (2016). Energy justice: A conceptual review. *Energy Research & Social Science, 11*, 174–182.

Jenkins, K., Sovacool, B. K., Mouter, N., Hacking, N., Burns, M.-K., & McCauley, D. (2021). The methodologies, geographies, and technologies of energy justice: A systematic and comprehensive review. *Environmental Research Letters, 16*, Article 043009.

Jerez, B., Garcés, I., & Torres, R. (2021). Lithium extractivism and water injustices in the Salar de Atacama, Chile: The colonial shadow of green electromobility. *Political Geography, 87*, Article 102382.

Kaushal, T., & Predmore, S. (2025). The women's inclusion agenda: Gender and everyday practices across registers of finance. *Finance and Society, 11*(1), 81–102.

Khasru, S. M., & Rafee, A. A. (2021). Renewable energy empowerment at the grassroots: The success story of solar home system (SHS) in Bangladesh. https://ic-sd.org/wp-content/uploads/2020/11/Syed-Munir-Khasru.pdf

Kingsbury, D. V. (2023). Lithium's buzz: Extractivism between booms in Bolivia, Argentina, and Chile. *Cultural Studies, 37*(4), 580–604.

Lacey-Barnacle, M., Robison, R., & Foulds, C. (2020). Energy justice in the developing world: A review of theoretical frameworks, key research themes and policy implications. *Energy for Sustainable Development, 55*, 122–138.

Langley, P. (2020). The financialization of life. In *The Routledge international handbook of financialization* (pp. 68–78). Routledge.

Lingham, J. T., & Johnston, M. (2025). Running on empty: Depletion and social reproduction in Myanmar and Sri Lanka. *Antipode, 57*(2), 494–514.

Magnani, N., & Osti, G. (2016). Does civil society matter? Challenges and strategies of grassroots initiatives in Italy's energy transition. *Energy Research & Social Science, 13*, 148–157.

Marx, K., & Engels, F. (1848). *The communist manifesto (1848).*

McCauley, D. A., Heffron, R. J., Stephan, H., & Jenkins, K. (2013). Advancing energy justice: The triumvirate of tenets. *International Energy Law Review, 32*(3), 107–110.

Murrey, A., & Mollett, S. (2023). Extraction is not a metaphor: Decolonial and Black Geographies against the gendered and embodied violence of extractive logics. *Transactions of the Institute of British Geographers, 48*(4), 761–780.

Navarro, R. (2022). Climate finance and neo-colonialism: Exposing hidden dynamics. *The political economy of climate finance: Lessons from international development* (pp. 179–203). Springer International Publishing.

Ojong, N. (2022). The dark side of the sun: Solar home systems and their injustices in Africa. In *Off-grid solar electrification in Africa: A critical perspective* (pp. 141–166).

Omstedt, M., & Ebner, N. (2024). Introduction: Uneven development and social difference in capitalism. *Environment and Planning a: Economy and Space, 56*(4), 1298–1303.

Ortiz, R. J. (2025). Spatial competition and unevenness in global capitalism: Labor, nature, and the seesaw of capital. *Critical Sociology, 51*(2), 283–300.

Pellicer-Sifres, V. (2020). Transformative energy transition from the bottom-up: Exploring the contribution of grassroots innovations in the Spanish context. *Innovation: The European Journal of Social Science Research, 33*(2), 124–139.

Rai, S. M., Hoskyns, C., & Thomas, D. (2014). Depletion: The cost of social reproduction. *International Feminist Journal of Politics, 16*(1), 86–105.

Ricci, A. (2021). *Value and unequal exchange in international trade: The geography of global capitalist exploitation.* Routledge.

Roitman, J. (2025). Financial platforms beyond the North–South divide. In *The Cambridge global handbook of financial infrastructure* (pp. 49–63).

Roitman, J., Moon, A., & Lin, L. (2025). Digital platform economies: Value from data?. *Platforms & Society*, 2. https://doi.org/10.1177/29768624251358643

Sareen, S., Sorman, A. H., Stock, R., Mahoney, K., & Girard, B. (2023). Solidaric solarities: Governance principles for transforming solar power relations. *Progress in Environmental Geography, 2*(3), 143–165.

Seyfang, G., & Haxeltine, A. (2012). Growing grassroots innovations: Exploring the role of community-based initiatives in governing sustainable energy transitions. *Environment and Planning c: Government and Policy, 30*(3), 381–400.

Smith, J. (2016). *Imperialism in the twenty-first century: Globalization, super-exploitation, and capitalism's final crisis.* NYU Press.

Smith, N. (2008). *Uneven development: Nature, capital, and the production of space.* University of Georgia Press.

Soederberg, S. (2020). *Urban displacements: Governing surplus and survival in global capitalism.* Routledge.

Sovacool, B. K., & Dworkin, M. H. (2014). *Global energy justice.* Cambridge University Press.

Sovacool, B. K., Hook, A., Martiskainen, M., Brock, A., & Turnheim, B. (2020). The decarbonisation divide: Contextualizing landscapes of low-carbon exploitation and toxicity in Africa. *Global Environmental Change, 60,* Article 102028.

Stock, R., & Birkenholtz, T. (2024). Tracking the sun: Exposing India's solar dispossessions. *Globalizations,* 1–20.

Stock, R., & Sovacool, B. K. (2023). Left in the dark: Colonial racial capitalism and solar energy transitions in India. *Energy Research & Social Science, 105,* Article 103285.

Vidal, M. (2019). Work and exploitation in capitalism. In *The Oxford handbook of Karl Marx* (p. 241).

Voskoboynik, D. M., & Andreucci, D. (2022). Greening extractivism: Environmental discourses and resource governance in the "Lithium Triangle." *Environment and Planning e: Nature and Space, 5*(2), 787–809.

Winther, T., Ulsrud, K., & Saini, A. (2018). Solar powered electricity access: Implications for women's empowerment in rural Kenya. *Energy Research & Social Science, 44,* 61–74.

Index

A
Africa
colonial electrification in 26
energy policy and aid dependence 72
solar finance and PAYG markets in 33
uneven development in 3
Agbogbloshie (Ghana)
e-waste recycling economy 165
gendered labor and health risks 11, 195
toxic afterlives of solar devices 147, 185, 186
Aid and concessional finance
concessional underwriting 39
multilateral institutions and market engineering 49
Results-Based Financing schemes 63
Aluminum 1, 27, 29–32, 34, 35, 87, 88, 147, 158
bauxite extraction and global solar value chains 147
Rio Tinto and Bosai Minerals operations 147
Assetization 134
datafication of households and energy use 134
securitization of micro-payment 134
Awaso (Ghana)
bauxite mining and labor conditions 9
dust pollution and health impacts 151, 152
global aluminum and solar supply chains 146
insertion and depletion dynamics 146, 147

N. Ojong, *Solar Power Capitalism*, Energy, Climate and the Environment,
https://doi.org/10.1007/978-3-032-13707-4

B

Bauxite mining
 Awaso mine 147, 151
 labor precarity and dust exposure 152
Bodies 3, 6, 14, 15, 133, 146, 147, 149, 157, 160, 163, 165, 168, 181, 185, 187, 189, 195, 204, 209, 212
 depletion of through labor and toxicity 160
 gendered exhaustion and care work 155
Bounded agency 78
 African governments and global constraints 77
Bracking, Sarah 4, 5, 205, 208

C

Capital, global 14, 49, 62, 137, 179
 concessional loans and aid regimes 62
 flows and frontiers of accumulation 211
Carbon neutrality and green transition 208, 213
 contradictions of decarbonization 212
China 45, 49, 69, 70, 72, 84, 88, 147, 161
 role in African solar expansion 71
 Suntech Power and global photovoltaic manufacturing 49
Colonial electrification
 enclave privilege and selective provision 26
 externalization of risk 28
 persistence in postcolonial energy systems 33
Communal land 11, 146, 148, 180–182, 205
 displacement and loss of subsistence 147
 enclosure for solar plants 148

D

Datafication 78
 digital footprints as collateral 136
 PAYG solar monitoring and surveillance 136
Debt and indebtedness
 household repayment ecologies 6
 securitization of microfinance 205
 women's moral responsibility for repayment 119
Depletion (core concept)
 bodily, ecological, and temporal exhaustion 3
 depletion as counterpart of insertion 3
 depletion of women's labor, care, and health 187
Development finance 59
 concessional and blended finance 62
 World Bank and IFC programs 9
Displacement and enclosure 151, 163, 180, 205
 gendered time tax 174
 Kaleo-Lawra solar plant 145, 209

E

Ecological degradation 4, 157, 165, 188

dust, fumes, and toxic residues 160, 209
mining and waste ecologies 28
Edea dam (Cameroon) 26, 30, 32
colonial hydroelectric infrastructure 26
Electrification, colonial and postcolonial 24, 26, 40, 190
design logics of exclusion 15
industrial and settler priorities 25
Enclave privilege (design logic) 23, 26, 33, 36, 37, 40, 146, 206, 213
concentration of power and access 165
persistence in PAYG markets 9
Energy justice 4, 24, 206, 211, 212
limitations of distributional and procedural frames 24
structural inequalities in access 58
Extraction 2, 7, 9, 29, 84, 87, 89, 129, 136, 147, 153, 185, 196, 205–209, 214–216
bauxite, labor, and global supply chains 2
gendered and ecological costs 137

F

Feminization of responsibility 186, 189, 195
moral economies of debt 108
women as guarantors of household repayment 174
Financialization
mobile money and micro-payments 10
PAYG solar systems and securitization 205
Results-Based Financing and concessional loans 50
venture capital and impact investment 62
Fintech
data monetization 110
digital infrastructures of enforcement 109
SIM-enabled lockout mechanisms 109

G

Gendered infrastructures
patriarchal governance in energy transitions 29, 77
toxic labor in scrap economies 189
women's time, credit, and bodily labor 196
Ghana
Agbogbloshie scrapyard 32
Awaso mine 147, 151
Kaleo-Lawra solar plant 11, 148, 180, 203
Green capitalism 2, 4, 13, 145, 149, 165, 206, 213
accumulation through insertion and depletion 213
philanthrocapitalism and moral markets 46, 55

H

Harvey, David 3, 5, 23, 47, 49, 147, 150, 207
spatial fix and accumulation by dispossession 23, 147
Household finance

gendered budgeting and credit pooling 107, 179
indigenous savings groups 108, 116, 126, 127
PAYG repayment regimes 117

I

Inclusion (financial, gendered) 3, 57, 107, 116, 175, 205
as mechanism of insertion 130
moral economies of inclusion 193
Infrastructure 11, 23, 24, 28, 29, 32, 33, 58, 66, 74, 82, 109, 116, 135, 145, 157, 164, 174, 182, 188, 208, 215
colonial grids and postcolonial plants 6, 15
gendered infrastructures of care and labor 179
Insertion (core concept)
disciplinary process that renders life investable 3
intertwined with depletion 79, 107
International Finance Corporation (IFC) 5, 54, 59–67, 72
role in solar finance and market creation 33

K

Kaleo-Lawra solar plant (Ghana)
displacement and fenced commons 168
enclosure of communal land 11, 205
gendered time tax 174
Kenya 1, 7, 9–12, 59, 65, 69, 70, 72, 114, 117, 132, 137, 212, 214
mobile-money repayment systems 9
PAYG solar home systems 205

L

Labor, gendered 11, 109, 181, 194, 195
reproduction of patriarchal hierarchies 177
women's unpaid and hazardous work 14, 159
Land enclosure 9
communal dispossession 181
infrastructural citizenship and exclusion 150

M

Malawi 9–12, 27, 37, 72, 77, 80, 92, 121, 122, 126, 129, 130, 204, 208, 214
women's financial decision-making and SHS adoption 176
Mbembe, Achille 145, 154, 156, 191
necropolitics and exposure to premature death 145, 191
Microfinance and financial inclusion 116
solar home systems and repayment obligations 116
Mining 9, 25, 27, 29, 32, 88, 151–153

bauxite extraction and ecological depletion 147
Mobile money 1, 10, 111, 134
digital payments and surveillance 138
securitization of data streams 110

N

Necropolitics 145, 158, 160, 191, 205, 207
calculated exposure to death in labor regimes 191
NGOs and philanthrocapitalism 58, 75
role in market-making and moral legitimation 58

P

Patriarchal governance 8, 13, 29, 33, 77, 138, 174, 175, 180, 204, 207, 209, 213
household hierarchies and gendered discipline 177
reproduction of patriarchal capitalism 209
Pay-as-you-go (PAYG) solar systems
digital repayment mechanisms 136
securitization of micro-debt 116, 210
SIM-enabled lockouts 109
women's responsibility and reputational risk 14, 176
Philanthrocapitalism 46, 55, 57
celebrity-driven electrification campaigns 57
humanitarian finance hybrids 46
Policy scaffolding 46, 50
aid-based de-risking in Africa 50
tax incentives and feed-in tariffs 52
Pollution 133, 154, 159–161, 164, 165, 207
dust, fumes, and waste ecologies 165

R

Rai, S.M. et al. 3, 182, 206
Reproductive labor 26, 108, 186
hidden subsidy of energy systems 4, 175
Results-Based Financing (RBF) 46, 50, 61, 63
technocratic de-risking and measurement regimes 46
Risk externalization 33, 40, 213
displacement of costs to local ecologies and bodies 28

S

Securitization 108, 125, 133, 135, 137, 205, 214
bundling of micro-debts 5, 116, 205, 210
circulation of risk through financial markets 71
Social reproduction 49, 185
unpaid labor sustaining green capitalism 11
Solar energy markets 9, 14, 46, 50, 54–56, 59, 62, 66, 73, 120, 174, 175, 206
engineering and policy design 46
global capital flows 14

Solar power capitalism (central concept)
- gendered and financialized structures 15
- global circuits of accumulation 208
- insertion and depletion as organizing logics 1, 4, 6, 9, 15, 157, 204

Structural violence 6, 145, 152, 153, 156, 158, 159, 187, 191, 205
- slow death and infrastructural neglect 191

Surplus capital 5, 49, 50, 93, 212
- redirection from petro-dollars to photovoltaics 48, 49

T

Tanzania 9–12, 25, 28, 38, 39, 54, 74, 75, 88, 111, 113, 122, 123, 127, 130, 132, 137, 173, 175–178, 204, 208, 211, 214
- colonial energy legacies 10
- PAYG systems and women's budgeting practices 9, 10

Toxicity and e-waste 2, 9, 14, 133, 145, 147, 148, 157, 158, 160, 165, 166, 185, 186, 190, 193, 204, 209, 213
- Agbogbloshie scrapyard 160
- gendered exposure and health depletion 159

Trade-offs (flour-or-light dilemma) 111, 113
- rationing and energy precarity 113

W

Women
- ambivalent empowerment narratives 8, 29
- depletion through care, credit, and time 185
- insertion into solar power capitalism 29

World Bank 5, 35, 38, 39, 54, 58–65, 67, 68, 71, 72, 75
- de-risking strategies and solar finance 59
- Results-Based Financing and concessional loans 58, 63

Z

Zones of sacrifice 7, 14, 211
- slow violence of green accumulation 156
- toxic landscapes and necropolitics 7

The manufacturer's authorised representative in the EU is Springer Nature Customer Service Centre GmbH, Europaplatz 3, 69115 Heidelberg, Germany. If you have any concerns regarding our products, please contact ProductSafety@springernature.com

Printed and bound by CPI Group (UK) Ltd, Croydon, CR0 4YY

17/09/2026

02218433-0003